A Splintered History of **WOOD**

• Spike Carlsen •

A Splintered History of **WOOD**

Belt Sander Races, Blind Woodworkers, and Baseball Bats

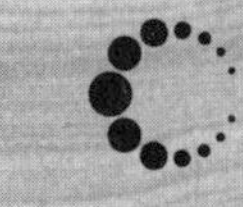

COLLINS
An Imprint of HarperCollins*Publishers*

HarperCollins books may be purchased for educational, business, or sales promotional use. For information, please write: Special Markets Department, HarperCollins Publishers, 10 East 53rd Street, New York, NY 10022.

FIRST EDITION

Designed by Nicola Ferguson

Library of Congress Cataloging-in-Publication Data
Carlsen, Spike, 1952–
A splintered history of wood : belt sander races, blind woodworkers, and baseball bats / Spike Carlsen.
p. cm.
Includes bibliographical references and index.
ISBN 978-0-06-137356-5
1. Wood. I. Title.
TA419C395 2008
620.1'2—dc22
2008001674

08 09 10 11 12 OV/RRD 10 9 8 7 6 5 4 3 2 1

For Kat: My xylem, my phloem, my roots, my sunshine,
the love of my life.

ACKNOWLEDGMENTS

Without a sliver of doubt, if it weren't for one extraordinary person—my wife, Kat—this book never would have been. She helped plant the seed for the idea, then warded off drought while it struggled and grew. I had wood on the brain for two years; she was always supportive. I went from full-time to half-time to zero-time at work to focus on the book; she never flinched. Her insights into the early drafts were invaluable. "Thank you" is not enough.

Thanks to those who took sometimes entire days out of their busy schedules to walk me through their worlds of wood, including Warren Albrecht; Jim Anderson and Paul Johnson of MaxBats; Elena Barinova; Liz Beiter; Thomas Boehm of Ancient Archery; President Jimmy Carter; Tom Caspar of *American Woodworker*; the staff at Cremona's International School for Violin Making; Livio De Marchi; Ken Felber; Sebastiano Giorgi; John Hughes, "The Mad Kiwi"; Russell Karasch of the Barrel Mill; John Kerschbaum; John Koster of the National Music Museum; Alan Lacer; Johannes Michelsen; Patrick Moore of Greenspirit; Mira Nakashima; James Olson of Olson Guitars; Dennis Roghair; Bob Teisberg of Ancientwood; Alex Weidenhoeft; interpreter Abbey Mahin; those who so generously allowed their photographs to be used; and others. I'd like to particularly thank wood identification and information specialist Regis Miller, formerly of the Forest Products Laboratory, for reviewing endless pages and answering endless questions.

Thanks to Alfredo Santana for getting the manuscript into the right hands and to Matthew Benjamin for being those right hands; to Dana Adkins, of the Adkins and Phillips Agency, whose admonition to "Get out of your chair and muck around more" breathed life into this project; and to Debbie Phillips for dotting the *i*'s. Thanks to the team at Collins—Bruce Nichols, Jean Marie Kelly, Stephanie Meyers, Kimberly Cowser, Helen Song, and especially my editor, Lisa Hacken, who let me say what I wanted to say, only better.

Thanks to my former colleagues at *The Family Handyman* magazine, especially Ken Collier and Mary Flanagan. Thanks to Frank Bawden, Dorothy Case, Tom Turnquist, and Troy Harper, each of whom provided his or her own unique brand of inspiration.

Thanks to Dan Poffenberger, Tom Thiets, Michele Hermansen, Marlene Harty, and those at Bomalang'ombe Secondary School in Tanzania for reminding me what's truly important.

Thanks to my remarkable father, who pointed me in the right direction (I miss you), and to my extraordinary mother, who kept me moving in the right direction. To my five wise and wonderful children—Sarah, Maggie, Zach, Kellie, and Tessa—you give me reason to live. To Paige—the future is yours; treat it wisely. And thanks to all the friends and family—you know who you are—who helped along the way.

CONTENTS

INTRODUCTION

When we think of wood—and few of us do—most of us picture the stacks of two-by-fours in the aisles of our local home center or the stuff we throw into the fireplace on cold winter nights. Wood doesn't rank much higher on our "things-that-amaze-us" list than water or air. We chop our onions on it, pick our teeth with it, pin our skivvies to the clothesline with it. Most people think of wood as just another "thing"—and they're correct.

But let's look at life for a minute without this thing. For starters, the book you are now reading wouldn't exist. If you needed to dab your eyes a bit over that fact, you wouldn't find a Kleenex or Kleenex box in the house. In fact, you wouldn't find the house, or the chair you are seated in or the floor it's standing on—at least not in the form to which you are accustomed. You wouldn't have the pencil in your pocket, the rubber heel on your shoe, or the cork you popped from the pinot noir last night. There would have been no violins at the concert you attended last week, no baseball bats at the ball game you watched last night, no telephone poles to carry your digital messages earlier today.

We use wood for chopsticks, bridges, and charcoal. From the cribs we sleep in as infants to the caskets in which we'll be buried in death, wood touches us in a real and personal way, every day. How could we take wood for granted?

And now I step off my soapbox—also made of wood.

If one thinks hard enough, one comes to realize that wood *is* a remarkable substance. And equally remarkable are the stories it has to tell. It's thrilling to run your hand across a polished tabletop in Bob Teisberg's showroom, but it's even more thrilling when you learn the slab of wood is fifty thousand years old, dug up from the peat bogs of New Zealand. The delight in running your fingers across a dovetailed cherry toy box built by Ron Faulkner is made more delightful by the knowledge that this woodworker is blind. The awe in watching a catapult hurl a pumpkin the length of a football field at the annual Punkin Chunkin Contest is made all the more awesome when one reads about the War Wolf, which, in 1300, could hurl stones weighing 300 pounds an equal distance.

In recent years a spate of books examining a single commodity has emerged. There are books on salt, dirt, dust, chocolate, clay, tobacco, ice, coal, cod, gold, and more. In every case, the author—as is the author's duty to do—makes it clear that without the subject at hand, the world today would not be as we know it. By the time you've finished reading *Cod: A Biography of the Fish That Changed the World,* you're convinced the United States would still be a British colony if it weren't for the billions of cod along America's coast. As you turn the last page of *Coal* you believe you'd be riding to work on horseback if it weren't for the black stuff. Skimming *Clay* makes you realize that the sticky stuff is responsible for everything from the glossy coating on the magazines you read to the toilet you sit on while reading them.

Though I've tried to temper my enthusiasm, this book joins the ranks of the commodity pitchmen. Without wood, it's not that we'd just be a little hungrier or a little bit more behind the times; it's that we—and I go out on a limb here—simply might not be here at all. We wouldn't have had the fire, heat, and shelter that allowed us to expand into the colder regions of the world. We wouldn't have had boats for exploring this wonderful planet. If every oxygen-generating, carbon dioxide–consuming, wood-producing tree on earth were to suddenly die, humankind would have a rough go of it indeed. The adage "Man has no older or deeper debt than that which he owes to trees and their wood" has a truthful ring to it.

Still, questions remain: Why do we continue to employ wood, even when cheaper, more durable materials are available? Why is it that, though we can create a dining room table out of carbon fiber that will never scratch, stain, or split, we still prefer to put up with scratch-able, stainable, splittable wood? Why is it that with electric pickups and sound effect modules that can create every sound in the book, we still choose the wood violin over its synthesized substitute? Why is it that Jimmy Carter, one of the busiest people on the planet, with the wherewithal to buy whatever furniture his heart desires, continues to craft cradles, tables, and chairs? Why is it that with steel studs that are lighter, cheaper, and less prone to fire and rot, we still build our houses of wood? This book is an attempt to find out.

Of course, this book didn't find out *everything*. Not even close. In fact, as the title of the book indicates, all we were really able to fit in were a few splinters of information. But we think we found some of the most interesting splinters.

Here is a look at wood and its splintered history.

A Splintered History of **WOOD**

CHAPTER 1

Extraordinary Woods

As I drive toward Ashland, Wisconsin, home of the company that lays claim to selling the oldest workable wood on the planet, the convoys of fully loaded pulpwood trucks I pass remind me of the rich, ongoing logging tradition of the area. I'm in Sawdustland. It's a fitting place for a company named Ancientwood to call home. I find the pole building that serves as the warehouse/store/Internet headquarters, and I find owner Bob Teisberg. He greets me by making three introductions. The first is to his shop helper, Dante; the second is to a mammoth slab of kauri wood standing by the door; the third is to his sense of humor. "Yep, we call that slab Dante's Inferno. He went through hell for two straight weeks sanding and finishing that baby. But just look at it." And when you look closely at this gigantic slab, you set your eyes on things of an unworldly nature. For starters, it's 5 feet wide, 7 feet tall, and 3 inches thick. It's sanded smooth as glass, with a finish and grain that not only glow but dance like a hologram, depending on your viewing angle. The color, figure, and texture are unlike any wood I've ever seen. And the reason is, it *is* a wood I've never seen. It's a wood most people have never seen. The slab is from a fifty-thousand-year-old kauri tree, mined from the bogs of New Zealand.

FIFTY-THOUSAND-YEAR-OLD WOOD LIVES AND BREATHES AGAIN

The route a slab of wood needs to travel to get from 48,000 BC on the North Island of New Zealand to AD 2006 in Ashland, Wisconsin, is not an easy, inexpensive, or clean one. "Originally we thought some cataclysmic event—a tsunami, an earthquake, an asteroid—was responsible for the death of the trees," explains Teisberg, the North American distributor for Ancient Kauri Kingdom wood. "But when they sent samples to the University of New Zealand for study, they found the trees died at different times and fell in different directions, so our best guess is they died of natural causes." But it doesn't matter so much how they died as where and when they died. When most trees die, they keel over and decompose within a few decades. But these kauri trees keeled over into bogs—an oxygen-starved, fungus-free environment—that created a time-warp cocoon that preserved the timber in pristine condition, until a Kiwi by the name of David Stewart happened along.

The Ancient Kauri Kingdom's informational DVD, in which Stewart stars, shows the process used to extract the trees. Most of the trees are found in farm pastures, where they reveal their presence by a small exposed section. "If you're a farmer you really don't want these things in your field," explains Teisberg. "Nothing grows on them, and animals can break a leg if they fall through a rot pocket, so they're just a nuisance." When they go into an area, they're never quite sure what condition or size the trees will be in; there's really nothing scientific about it. They get in there with a backhoe, give the exposed part a wiggle, and if the land 100 feet around them moves they know they've got a monster. And they've found some monsters.

The extraction process involves moving man and machine across the boggy land, trenching all around the log, then using a chainsaw with a bar the length and lethalness of an alligator to cut the log in two if it's too large to get out in one piece. The video of the process, which absolutely oozes testosterone, shows a cigarette-chomping Stewart, covered in slime, standing in the bucket of the backhoe, sawing a 60,000-pound monster in two with a chainsaw sporting a 6-foot-long bar. There are hydraulics, chains, cables, muck, and heavy machinery everywhere. The wood chips flying out of the kerf look as clean and uniform as if he were slicing through a 25-year-old birch tree. At one point he pauses to show the camera a handful of forty-five-thousand-year-old kauri leaves.

Once the sections are cut to manageable size, they're winched, pushed and pulled up out of the trench, rolled onto massive flatbed trucks, and then hauled to the company's yard, where they're marked and cut into slabs. The logs have reached the 100 percent saturation point after lying in the bogs for eons, and the drying process is a long drawn-out affair as the wood finds a new moisture balance.

The crown prince of kauri logs is the 140-ton "Staircase" log discovered in October of 1994; the largest known log of any kind ever to have been extracted anywhere. The crew broke two 90-ton-capacity winch cables attempting to haul the trunk out in a single piece. They cut the tree into separate 110- and 30-ton sections, hauled the sections out, and then let them sit untouched, not wanting to cut the trunk into slabs because of its Olympic-caliber size. Four years later, Stewart built a 20-inch-thick reinforced concrete pad, placed a 50-ton, 12-foot-diameter, 17-foot-tall section of log on top of it, and went after it with a chainsaw. After three hundred hours of carving and two hundred hours of finish work, the world's largest, and surely oldest, single-piece circular stairway was complete. It's built *inside* the log. If you pause to count the growth rings as you're ascending you'll find 1,087 of them.

The scene in Ashland, Wisconsin, is considerably tamer. Teisberg walks me past pile after pile and specimen after specimen of imported ancient kauri. He has everything ranging from 6-foot-thick stumps to 1/16-inch-thick veneers. At one point, Teisberg stocked what he claimed to be the "largest single piece of wood available in the United States"—and I never found any challengers. The slab measured over 20 feet long,

A slab of fifty-thousand-year-old kauri wood, 20 feet long and 5 1/2 feet wide, claimed to be the largest single piece of wood available in the United States. The slab, 4 1/2 inches thick, contains over 500 board feet of wood and zero knots.

5 1/2 feet wide, and 4 1/2 inches thick; it was estimated to have grown for a thousand years, and, amazingly, it contained not a single knot.

Kauri sells for $35 per board foot, a price comparable to that of high-grade teak today. "Teak is beautiful," explains Teisberg. "But you'll find it on every freakin' sailboat made today. If you dig the [kauri] story and you want something exotic, then you're way in. If not, head to Home Depot."

It's not only boatbuilders who dig the kauri story. Scientists are studying the growth rings to get a read on the climate and environment fifty thousand years ago. Many of the boards have fifty to sixty growth rings per inch. They have stories to tell.

Fifty thousand years old is getting on in age for a piece of wood, but Mike Peterson, a forester with Forestry Tasmania, believes he's found wood that makes ancient kauri wood look downright pubescent. In the 1930s, Huon pine logs that had been buried in an alluvial plain in the Stanley River region were uncovered during a tin mining operation. Initially pieces were dated as being 7,500 years old. Then, in 1994, carbon dating revealed some to be 38,000 years old. Now scientists are examining Huon pine logs containing extraordinarily wide growth rings, revealing that these trees grew during an exceed-

A massive kauri log being extracted from a bog in New Zealand and loaded on a flatbed trailer with the help of two backhoes and a bulldozer. It's not unusual for a log of this type to weigh 30 tons and to be 15 feet in diameter.

ingly warm period of the earth's history—perhaps preglacial—which could make them 130,000 years old. But the tree ring chronology jury is still out.[1]

The oldest nonpetrified piece of wood on the planet may be the small hunk of Cupressinoxylon wood that wood collector Richard Crow has sitting on a shelf. It's estimated to be seven million years old but, apart from its deep black color, "looks like it was felled a year ago," according to Crow.

None of this makes kauri wood any less amazing. Though the wood looks fairly unremarkable in its raw state, it begins emitting its trademark opalescent glow once sanded down to 1200 grit and given a finish. The farther down you get into the base of the trunk and root area, the wilder the grain and figure become. Furniture builders love the stuff, crafting it into both highly machined and natural-edge pieces. Turners like turning it wet, letting it dry out—sometimes for as long as two years—then turning it again to final shape and thickness. Musical instrument builders—including those who make guitars, ukuleles, drums, flutes, and harps—love both the look and sound. One acoustic guitar maker now uses kauri exclusively, and electric guitar builders in particular go wild for the boards with the wildest grain. One woodworker/jeweler sells rings turned from ancient kauri and touts them as keepsakes that connect the wearer to their prehistoric past.

The wildest of the wildest grains is called white bait, named after schools of small fish near New Zealand that emit an iridescent glow when swimming in one direction, then seem to disappear when changing course. "People ask me to describe white bait over the phone, and it really defies description," explains Teisberg. "There's no short description; it's like a confluence of grain activities. I've never seen it in another type of wood. You just sort of have to see it." And when you hold a sanded, polished, and finished board of white bait in the sun, you see what he means. It has depth; it shimmers; it plays practical jokes on you, depending on how you turn it.

As we head back toward his office Teisberg picks up a slab of kauri and tells me to rub my thumb "until it gets hot" over an area of the bark that contains an amber-color residue. "Now close your eyes and smell your thumb," he says. "That's what it smells like to stand in a fifty-thousand-year-old forest." He may be right, but the odor is so intense that I feel as if I'm standing in a 55-gallon drum of turpentine. The residue on my thumb is the dried sap of the kauri, which clings to the bark whether the tree is long dead or still growing. In the not too distant past this sap was collected, purified, and sent by the boatload to England and Norway to make linoleum and varnish.

Kauri was to New Zealand what white pines were to North America: massive and abundant trees ready for the taking by early European settlers, who harvested them for houses, shipbuilding, furniture, and firewood. When Captain Cook first reported the existence of "the finest timber my eyes have ever seen" in 1769, kauri forest blanketed about 4 million acres of New Zealand's North Island.[2] The trees were massive by any standards. If you were a European carpenter, a single kauri could provide enough wood for six houses. If you were a Maori warrior, you could craft one into a 115-foot-long war canoe, capable of carrying a crew of eighty.

Some monsters escaped the guillotine. Tane Mahuta, perhaps the largest kauri still growing today, measures 45 feet in circumference and stands 170 feet tall. But a tree known as Kairaru, which was destroyed by fire in the 1880s, was three times the size of Tane Mahuta. When living it was the largest tree by volume in the world, larger than the largest redwoods today, and was estimated to be over four thousand years old.[3]

Before leaving, I decide to purchase a free-form slab 16 by 24 inches and 3 inches thick, sliced from the base of an ancient kauri. I gulp a bit when Teisberg calculates the board feet and the total comes to $315. But it's a gorgeously entangled slab and, like a fine art lithograph, comes with its own serial number and certificate of authenticity—a certificate that reads in part:

> *This prehistoric kauri timber is from the forests buried during the 1st Ice Age, which are located on the Northern Island of New Zealand in the South Pacific Ocean. Our company, Ancientwood, Ltd. is satisfied that extensive and conclusive independent Radio Carbon Dating tests verify this age beyond doubt.*

And I think, "For a slab of fifty-thousand-year-old wood that's traveled halfway around the world, 315 bucks is a pretty good deal."

IN QUEST OF THE WORLD'S MOST EXPENSIVE BOARD FOOT

Hardwood lumber in the United States and Canada is sold by the board foot—a theoretical piece of wood 1 inch thick, 12 inches wide, and 12 inches long. The boards at your local lumber supplier may be any thickness, size, or shape, but when it comes time to tally up how much wood is stacked in the back of your truck or tied to the roof of your car, the formula is this: thickness × width (in inches) × length (in feet) divided by 12. That number is next multiplied by the cost per board foot of the wood you've selected: a price that can range anywhere from under a dollar for pine up to—well, just what is the upper limit? I went to find out.

If you're in search of the world's most expensive board foot of lumber, you start at the top: you talk to the King of Cocobolo. But when you locate him, you don't find an exotic, velvet-clad man from some remote Central American country. You find a guy in blue jeans, tennis shoes, and a ratty sweater by the name of Mitch Talcove in a dusty shop in Carlsbad, California. His company, Tropical Exotic Hardwoods, has been importing hardwoods from Mexico and other parts of the world since the 1970s, and after all these years he still admits, "Just when you think you've seen it all, something will come in hidden in a containerful of logs and you'll think 'Oh my God, nature is messing with my head again.'"

His namesake wood—cocobolo—is a majestic wood, with the heartwood ranging in color from an imperial orange to a royal red, and a strength that rules the charts in nearly every category. It's put to majestic uses, often turned, carved, sculpted, and inlaid. Much of it winds up as cutlery handles, since its density makes it capable of standing up to nearly all forms of culinary abuse, and its natural oiliness allows it to be soaked, washed, and rinsed eternally without losing its regal stature.

The King of Cocobolo handles more than cocobolo. Talcove explains that some woods he carries, like snakewood and pink ivory, are rare, exotic, and expensive, but they *are* commercially available. The rarest woods are those for which there's no regular source: woods like chittamwood or smoketree burl from the Deep South, which often grows interwoven among granite boulders and must sometimes be dynamited out. "You never know when it's going to be available," says Mitch. "It's a gemstone wood." A gemstone that can cost $35 per pound.

His most expensive piece of wood? Today it's a slab of true Cuban mahogany that's 2 inches thick, 2 feet wide, and 12 feet long, endowed with a mesmerizing ribbon grain. The tree was uprooted when Hurricane Hugo hit the Carribean. The King of Cocobolo has turned down $10,000—slightly over $200 per board foot. If you want it, expect to pay a king's ransom.

Sam Talarico gives this slab of figured claro walnut crotch, just opened on his bandsaw, two thumbs up. The wood will eventually sell for up to $100 per board foot.

There are other kings in the world of exotic lumber, and Sam Talarico of Talarico Hardwoods in Mohton, Pennsylvania, is one of them. On his Web site's "Wood Porn" section he explains his passion: "There is nothing to compare with the feeling and excitement of opening up a highly figured log and seeing what's inside. We do this every year and I want to share some of these intense moments and very special figured lumber with all my loyal customers and all of you out there who are simply turned on by great wood. We choose to call it WOOD PORN which it certainly is to those of us that get the fever when looking at fantastic wood." Scroll through the photos and you find impossible woods: A slab of curly English walnut the size of a school bus, Volkswagen-size crotches cut from Cairo walnut, a 400-year-old English Oak log that's 30 feet long and knotless (or knot-free).

He specializes in woods from England, Scotland, France, Germany, Russia, and other parts of Western Europe. He travels, sleuthing out the most spectacular logs, buying them, fumigating them as required by law, and wrangling with the myriads of regulations and paperwork before shipping them via container to his yard. Sam is the master of ceremonies when it comes to opening each log, personally studying, plotting, and marking each before committing it to the saw. He compares the process to cutting diamonds. He uses a restored Dolmar saw he found in the weeds behind a sawmill in England. It's a gigantic bandsaw affair, powered by hydraulics and a diesel engine capable of cutting slabs up to 8 feet in diameter.

After thirty-five years in the business, Sam knows exactly which trees are diamonds in the rough and which are saw blade killers. He avoids those growing along fencerows or in backyards, which are more likely to contain nails, horseshoes, metal posts, and cement. "One time I found an entire axle from an old wagon inside a log," he relates. "Someone must have leaned it again a tree two hundred years ago, and the tree grew around it."

He's Lumberman to the Stars, having supplied lumber for furniture built for Tom Hanks, Rene Russo, Charles Schwab, and others whose names he can't reveal because of nondisclosure agreements he's had to sign. He reminisces about "the perfect oak log" he found in West Virginia in the 1970s: 4 feet in diameter, 60 growth rings per inch, flawless. "Thirty years later, and people still talk about that lumber; it was absolutely perfect." And what's the rarest, most expensive board foot of wood in Sam's vault? It's the highly figured wood from a curly English walnut log he purchased from a Mennonite farmer several years ago. The price tag: $250 per board foot.

I wind down my search by chatting with Rick Hearne of Hearne Hardwoods. There may be more expensive wood somewhere, but when you find the guy who has hauled koa logs from the jungles of Hawaii using helicopters, has cut seven-hundred-year-old burr oak from England's Sherwood Forest, and stocks over 1 million board feet of lumber ranging from African anigre to Guatemalan ziricote, you figure the end of the quest must at least be near.

Hearne stocks amboyna burl—a wood of intense beauty and depth created by a "cancer" that infects the tree. "If you were to talk to exotic wood dealers around the world, this would be on the short list of the five most exotic woods in the world." His largest specimen—

a 275-pound slab 3 1/2 inches thick, 42 inches by 48 inches—will set you back $110 per board foot, or a total of $5,000, but still not close to his most expensive offering.

His ancient bog oak—a wood that's chocolate on the outside and sunburst on the inside—is another rare offering. In the 1800s a reservoir in Austria was built and the area was flooded. Five years ago the reservoir was drained, and while it was being dredged deeper, white oak trees were found that have since been carbon dated by the University of Salzburg as being forty-five hundred to five thousand years old. Buying one is a game of chance. You tell them how many you want, they bring in a crane, and you buy whatever emerges. Rick has never been disappointed.

Hearne knows about big. At the time we spoke, he was awaiting delivery of a slab of sapele wood from Africa—5 feet wide, 25 feet long, and 3 inches thick—for a client in need of a rather large table.

He bemoans the fact that good saw logs are increasingly difficult to find in the United States. Few large-scale efforts, public or private, are being made to replant cherry, walnut, and other hardwoods for the woodworkers who will be crafting fine furniture two hundred years from now. "But," Hearne explains, "in Germany they don't talk about managing a forest; they talk about building a forest. One forest there has been managed since 1720. Trees are harvested on three-hundred-year cycles, which means 1/300th of a forest is cut per year. North American plans are based more on thirty- to sixty-year cycles."

Logging in Europe is not without its hazards. In areas where trench warfare raged during World War I, the mills carry shrapnel insurance. "A single piece of hardened shrapnel in the mild steel rollers of your bandsaw mill will totally destroy them," Hearne says. Along the same lines, he talks of a walnut tree he cut in Westchester, Pennsylvania. "It was out in the middle of a woodlot with no fences around and no reason for anyone to drive a nail into it. But it turns out a previous owner had owned a 50-caliber machine gun and used the tree for target practice. The tree was absolutely loaded with 50-caliber bullets." Bullets that didn't help his saw any.

When asked if he's a woodworker himself, Hearne explains that he's an okay woodworker, but with customers like Sam Krenov, Sam Maloof, and Wendell Castle—superstars of the woodworking world—he's surely hesitant to call himself a great one.

So what's the most expensive board foot of wood this "okay woodworker" carries? Rosewood burl: $350 a board foot. At that price, wood to make a 1-inch-thick top for a standard 3-foot by 3-foot card table would sit at $3,150.[4]

OAK: THE BREAKFAST OF CIVILIZATIONS

When I started researching this book, I vowed not to use the blanket phrase "No other wood/tree/woodworker has played a greater role in the history of mankind than _____." But, damn, I came close with oak.

Though perhaps a bit overzealous in his admiration of the species, William Bryant Logan, in *Oak: The Frame of Civilization,* states:

> For ten thousand years—oak was the prime resource of what was to become the Western World. Through Dru-Wid, "oak knowledge," humans learned to make homes and roads, ships and shoes, settles and bedsteads, harness and reins, wagons and plows, pants and tunics, swords and ink.[5]

Without question, oak proved itself to be an indispensable companion as civilization became, well, more civilized. Because of its strength it was the preferred material for building the ships that explored the New World. Because it was easily split and long lasting, it was used for fencing, which helped domesticate animals. Because of its denseness and easy workability, it was used for the gears of the earliest machines—windmills, waterwheels, clocks, and mills. It was used for barrels, which transported bulk items for trade and consumption. It was used for furniture, roads, heat, and buildings.

The bark was used for tanning leather. Decoctions of inner bark were used to treat sore throats, ulcers, hemorrhoids, and sore eyes; indeed, it was listed in the *United States Pharmacopeia* as a recognized drug until 1936.[6] Oak galls, a reaction to a parasitic wasp, were used for creating ink.

Charcoal was the fuel that powered most of man's early industrial efforts, and there was no finer wood than oak, with its high heat content, from which to make this charcoal. Brickmakers, glassblowers, ceramists, and iron makers all used prodigious amounts of oak charcoal. It was used for refining sugar, boiling soap, and burning the lime required for mortar. It was the natural gas of the classical, medieval, and Renaissance worlds.

Wolfgang Puck's forerunners were quite creative when it came to using acorns for cooking. Balanocultures—cultures that have relied heavily on acorn consumption for survival—have been found worldwide and throughout history. As early as the eighth century BC, the Greek poet Hesiod wrote: "Honest people do not suffer from famine, since the gods give them abundant subsistence: acorn bearing oaks, honey, and sheep" (though, we hope, not all in the same dish). Ovid, Lucretius, and Pliny all mention acorns as a splendid food source. In Tunisia, the old word for oak translates as "the meal-bearing tree."[7] One study hypothesizes that it took one-tenth the time to harvest acorns than it did to harvest wheat or barley.[8]

The recipe book is thicker than one might suspect. The Chinese still whip up a wicked acorn stew, the Turks a hot acorn-based drink called *racahout,* the Spaniards an acorn liqueur and olive oil substitute.

One California-focused study concluded that acorns could have fed Native American villages of up to a thousand people and that two to three years worth of acorns could be gathered and stored in just a few weeks—not surprising, given that a single large oak can bear up to 500 pounds of acorns. (Oaks *need* to be prolific, since less than one out of every ten thousand acorns becomes a tree, and many of those won't start producing acorns until they're fifty years old.) Sue Ellen Ocean, who lives in Willits, California, has recently published a cookbook, *Acorns and Eat 'Em,* which contains recipes for acorn cereal, acorn pancakes, acorn lasagne, and acorn enchiladas. While

Cork is harvested from evergreen cork oaks on seven- to nine-year cycles, with each tree yielding around 400 pounds of cork per stripping—enough to make twenty-five thousand wine bottle corks.

most acorn concoctions are reportedly bland in taste, they make up for this by being exceptionally filling.

Oak produces another surprising progeny. Cork is harvested from the bark of the evergreen cork oak, which grows primarily in Portugal, Spain, southern France, Italy, and North Africa. People have used cork for five thousand years for items ranging from simple floats used by Chinese fishermen to sandals worn by ancient Greeks. Cork can last decades, even centuries, as witnessed by the seventy-year-old cork floors found at the St. Paul Public Library. It's excellent for flooring because it's elastic and flexible, sound-absorbing, and warm underfoot.

The cork oak doesn't require herbicides, fertilizers, or irrigation, and it regenerates itself after each harvesting of the bark. After the tree has reached the age of twenty years, cork is normally harvested on seven- to nine-year cycles.[9] Over the life span of a tree—one hundred fifty to two hundred years—each cork oak can provide fifteen or more strippings. While most cork oaks yield around 400 pounds of cork per harvest (around 25,000 wine corks' worth), the world's record is 3,870 pounds, stripped from a single tree in Portugal in 1889.[10] Trees in general are considered a fabulous renewable resource; in this light, cork oak ranks as sublime.

There are somewhere between two hundred and four hundred fifty species of oak, about half of them being evergreen. They're distributed primarily about the northern hemisphere and can be found in locations as diverse as the Scandinavian countries, Sicily, and Japan. We love oak so much that it is the official national tree of the United States, Germany, and Great Britain. It's the species of tree that is struck by lightning the most, and some surmise that its tendency to burst into flames on such occasions has something to do with some early civilizations' inclination to impart mystical powers to the tree.[11]

Oak is a "Mama Bear" kind of wood: not too hard and not too soft; not too stiff, yet not too pliable; the grain falls somewhere between fine and coarse, it has interesting grain patterns, but the grain is not so interesting as to be distracting. In his classic book *A Natural History of North American Trees,* Donald Peattie eloquently waxes, "True, White Pine warps and checks less, hickory is more resilient, Ironwood is stronger and Locust more durable; but White Oak would stand second to almost all these trees in each property in which they excel, and, combining all these good qualities in a single species, it comes out in the end as the incomparable wood for nearly every purpose for which wood can be used."

White oak is so valued throughout Europe that oak forests are managed, cultivated, and handed down through families for centuries. Oaks are planted in cycles lasting a hundred and fifty years or more. To shade the trunks to inhibit side branching—thus creating more valuable, clear, knot-free lumber—beech trees are planted surrounding each oak. Three or four generations of beech trees are harvested to provide income during the time it takes an oak to mature, but each oak is allowed to reach its prime. Patrick Moore, an environmentalist and third-generation logger, explains, "Forestry is one of the few industries where altruism is a requirement, because four or five generations of people may tend a tree knowing they will never directly benefit from it. By caring for a two-hundred-year-old oak, they're both paying respect to their ancestors and providing for future generations."[12]

But a good oak is worth the effort. It's the indisputable champion when it comes to making ships, furniture, flooring, barrels, cabinets, charcoal, moldings, and a hundred other products. It is so versatile that its name crops up in nearly every chapter of this book. It's indeed the breakfast of civilization.

THE WOOD FREAK SHOW

Snow forms in so many guises that the Inuit have twenty names for it; *karakartanaq* is "crusty snow that breaks under foot," *upsik* is "wind-beaten snow," and *qali* is "snow on the boughs of trees." Wood is equally eccentric; it's all over the map when it comes to looks, workability, strength, odor, texture, and other qualities.

One characteristic that makes wood unique is that even a single piece of wood is variable unto itself. If you take a cubic foot of most things—water, plastic, iron, Jell-O, Styrofoam, or granite—place it in a vise, and squeeze, it will react the same way no matter which sides of the cube are between the jaws. These objects are isotropic (they have identical properties in all directions) and homogenous (they're uniform in composition). But not wood; wood is anisotropic and heterogeneous. Depending on which way you place it in the vise—or drill it, dry it, stretch it, glue it, screw it, plane it, cut it, or almost-anything it—it will react differently. It even looks different from surface to surface. A cube of oak may be a bull's-eye of concentric circles on one surface, a bevy of lines on another, and a blank slate on yet a third.

If it's not challenging enough to sort out the differences within a single piece of wood, then head to your local lumber dealer or hardwoods store. There you can be amused by an entire freak show of woods—a display that includes the arboreal counterparts of the fat man, Leopard Girl, and Tom Thumb.

Weight and density. There are over a hundred species of trees and shrubs in the world with wood so heavy that they'll sink. *Specific gravity* is a ratio used to compare the weight of oven-dried wood with that of an equal volume of water. A cubic foot of water weighs 62 1/2 pounds. If a wood weighs 31 1/4 pounds per cubic foot (as black cherry does), its specific gravity is expressed as 0.5. The heaviest of the heavyweights are certain tropical ironwoods (a generic, not a scientific, name) with a specific gravity of 1.49 and a weight of 93 pounds per cubic foot. The heaviest and densest of these ironwoods are often referred to as *quebracho,* which fittingly translates into "axe-breaker." Some of these woods are so dense they've been used for anvils. The wood of the canyon live oak was so invincible that it was used by early pioneers for crafting both splitting wedges and the mauls that whacked them. Some of the hardest hardwoods register 2.5 on the mineral hardness scale; copper ranks 3.5. Hardness is closely related to density, which is closely related to weight. If you've got heavy wood, you've got hard wood.

The lightest of the lightweights is the Cuban wood *Aeschynomene hispida,* with a specific gravity of 0.044 and a weight of just under 3 pounds per cubic foot. If you wanted to balance a scale with a cubic foot of ironwood on one side, you'd need to place thirty times as much of the Cuban wood on the opposite platform. Balsa is the lightest commercially used timber; its very meaning in Spanish is "raft." It's used for projects as varied as aircraft, buoys, insulation, and theater props. The tree grows fast, dies young, and lives wet. Its moisture content is typically in the 300 percent range, and there are reported instances of it's approaching 800 percent. The trees are ready for harvesting by the age of seven and begin rotting in their early teens unless harvested and dried.

Regis Miller, retired wood anatomist with the Forest Products Laboratory, shows off an early two-finger bowling ball crafted from solid lignum vitae.

Color. The name Roy G. Biv should ring a bell—at least if you were paying attention to the mnemonic taught in grade school to help memorize the colors of the visible spectrum: red, orange, yellow, green, blue, indigo, violet. One can come close to creating this rainbow of colors with woods of the natural world.

Red you could glean from the redwood or incense cedar. Orange you could pluck from the Osage orange or yew. Yellow could be whittled from the yellow poplar or yellow buckeye. Green could be shaved from the magnolia. Blue is a rarity in woods in their natural state, though you could harvest plenty of it from lodgepole pines that have been infected by the mountain pine beetle. (The sapwood, colored by a blue-staining fungus, is even marketed by one company as Denim Wood.) Indigo and violet could be requisitioned from purpleheart. If one wished to add the color that represents all the colors of the spectrum, one would need to add holly for white. And if one wished to represent no color at all, one would select black ebony.

It is extractives—the chemicals that work their way into the heartwood belly of the tree as it ages—that produce the richest colors. Different genetic traits and environmental factors produce different extractives, which produce different colors in different woods. Some colorations are quite distinctive and specific; some lumber buyers specializing in African mahogany can reportedly tell which area and port a specific log came from, on the basis of differences in color in the heartwood.[13] In some woods the transition from sapwood to heartwood occurs in subtle gradations, but in others it is startlingly abrupt. A cocobolo board I have perched on my desk has white sapwood that butts up to the deep red-brown heartwood with absolutely zero hemming and hawing in between. And colors aren't stagnant; anyone who's left a hammer on a piece of freshly cut cherry and later returned to find a silhouette of that hammer will attest to that. Purpleheart morphs from purple to brown when exposed to light and air; eastern red cedar can do just the opposite.

Figure. "Figure," in simple terms, is "good grain gone wild." A well-behaved tree in a well-behaved environment will normally produce nice consistent growth rings. When that tree is sawn into boards, those growth rings produce grain and figure that's straight, consistent, and predictable. Figure—and when most woodworkers use this term, they're referring to *pronounced* figure—is an aberration in this consistency. For a slab of wood to be branded as figured, three factors come into play: (1) the type of aberration, (2) the way the aberrant section of wood is cut, and (3) whether one perceives it as beautifully figured or something defective to be thrown on the scrap pile.

One hallmark of most figured woods is a three-dimensionality and translucency. You'll hear terms like "shimmer," "depth," and "luster" bantered about in the hardwoods store. Most of the monikers used to describe figure are quite descriptive. Blistered, ribbon stripe, snail quilt, and lace figure are a few of the less common types of figure. Here are a few of the more common.

Bird's-eye figure is (surprise) a grain pattern reminiscent of hundreds of small bird's eyes scattered across the face of a board or veneer. It's created when growth rings are distorted as if they'd been poked by a dull pencil. Exactly *what* this dull pencil is in nature is unknown. Some maintain it's actually caused by birds pecking on the tree; others claim the cause is a type of mutant bud that attempts to sprout within the tree, rather than outside of it. Fungi, soil conditions, stunted growth, and other causes have been suggested, but none have been proved.

It's most common in hard maple but also occurs in birch, ash, and, in rare cases, black walnut. Michael Snyder, a forester in Vermont where maple reigns supreme, explains: "The bird's-eye pattern in wood is much like maple sugaring and fall foliage. We're familiar with it, we value it, and we know a tremendous amount about it. And yet, ultimately, it remains a mystery. For the better part of a century, wood scientists have been beating up on each other's explanations of what causes it. Hell, there isn't even agreement on how to spell it."[14] Snyder has discovered one thing characteristic of most bird's-eye-bearing maple trees: a Coke-bottle shape to the lower part of the trunk.

Spalted wood is wood that has been infiltrated by "waves" of decay, with each wave leaving a uniquely outlined stain-zone line. The look is not unlike that of the amoebalike figures projected over the stage during a 1960s Grateful Dead light show. Temperature, humidity, type of fungi, and chemical reactions all affect the end result. Once the wood is kiln-dried, fungi can no longer grow, and the spalting becomes frozen in time. The key is to catch the wood after the magic has begun but before it gets too punky. Woodturner extraordinaire Alan Lacer explains: "When wood is captured somewhere between the extremes of being completely sound and fully rotten, it can display magnificent beauty."[15] Again, hard maple seems to be the most frequent beneficiary of this beauty, and lighter woods in general offer the best canvas for Mother Nature to show off her flair for contemporary art.

Lacer offers his recipe for those wishing to create their own spalted wood. Place a freshly cut 2- to 3-foot long log upright on the bare ground, place a mound of dirt on the top end, and cover it loosely with black plastic. Keep it at a temperature between 60 and 80 degrees F. When the right amount of spalt has been attained, lower the humidity level to stop the progress then dry it. Since your next step is to cut, turn, or route this fungi-laden wood, it's highly recommended that you wear a respirator—especially if you have allergies or a weakened immune system.

Burls are the geodes of the woodworking world: baffling in their creation, plain-Jane or outright ugly on the outside, but often magnificent when cut open to reveal the mystery within. They're so convoluted and measurement defying that they're often sold by the pound instead of by the board foot. Burls are often described as a cancerous growth—and this description may not be too far from the truth. Most appear to be some type of genetic flaw that manifests itself in the form of a knoblike outgrowth. They can occur on any tree, anywhere, but are commonly found on elm, walnut, cherry, redwood, oak, and (again) our old friend and free-spirited maple. On the basis of artifacts, kings and queens of bygone days seem to have been particularly fond of burlwood items. Really, the ugliest thing about burls may be their alleged involvement in introducing Dutch elm disease into the United States; the disease may have come over from France in the 1920s when elm burl was being imported to create veneer for the furniture industry.[16]

Quilted figure, like quilts themselves, comes in many patterns. There's cloud quilt, tube quilt, bubble quilt, muscle quilt, and, for those who prefer to visualize in gastronomic terms, popcorn and sausage quilt. Quilted figure is revealed when wood with wavy grain is flat sawn. The overall effect is a surface with a soft, cumulus cloud–like appearance. Again, it is a maple—this time, bigleaf maple—in which this pattern most frequently manifests itself. *Blister figure* is the miniaturized form of quilted figure, and *quittle* is the nickname bestowed upon wood that has a blend of both quilted and curly figure. When cabinetmakers and wood turners want to create a piece with depth and striking figure, they often reach for their stash of quilted and quittled maple.

Wavy, ribbon, and *curly grain* are by-products of spiral grain that reverses itself periodically as a tree grows to produce something called *interlocked grain.* Visually these boards have a washboard effect, and as the varied grain intersects the wood surface and light at different angles, a hologram-type of depth emerges. *Fiddleback* is the term often used to describe wavy grain in maple, since it's frequently used for—you guessed it—fiddles. And the list goes on. *Bear scratch figure* forms when growth rings are indented. *Bee's-wing mottled figure* is a confluence of several grain patterns. *Crotch figure* reveals itself when the crotch or branch of a tree is cut lengthwise. There's *angel step, cat's paw, peanut shell,* and *flower grain.* And it is this rich, rare, one-of-a-kind figure that is the golden grail from which woodworkers can create works of unspeakable beauty—and for which hardwood dealers can charge unspeakable prices.

As much as we love figured wood, we really know very little about how it's formed. Scientific study is lacking; trees do not fit in test tubes, and rare is the scientist who has the patience, foresight, and funding to conduct a thirty-year-long experiment, with little chance of monitoring progress along the way, and no way of determining results without a chainsaw. But it seems there's little we can do to create it or even encourage figure. It doesn't seem to be affected all that much by climate, soil conditions, geographical location, growth rate, tree size, or disease.[17] We do know about figure formation in one specific tree: when strangler figs wrap their vines tightly around young mahogany trees, the resultant struggle gives birth to *drape* figure.

There have been a few successful attempts at growing or stimulating figure. A blistered or crinkle figure can be "manufactured" in Japanese cedar by propagating cuttings taken from older trees with crinkle figure. There's been success in creating figure in the same tree by binding bamboo sticks tightly around the trunk with elastic cord to create "dents" in the growth rings. There's been limited success in creating burls in boxwood by putting close-fitting metal bands around stems. Burls can also be coaxed into African thuja by repeatedly burning certain areas of the trunk and branches.[18] In 1929, J. F. Wilkinson grafted cuttings from a figured walnut tree and saw evidence in the offspring twenty-two years later. But for the most part, figure remains a mystery. It may simply be that each tree, like each person, is an individual and has a personality that's a product of both nature and nurture.

Odor. Wood knocks on all of our senses, including smell. Blindfolded, nearly everyone can detect incense cedar; average woodworkers can often sniff out sassafras, red cedar, and Douglas fir; and avid woodworkers with a keen nose can detect the subtle aroma of catalpa, teak, and other woods they commonly work.

Cedar is perhaps the most commonly harnessed wood fragrance today. You can find it in board form as cedar closet lining, in stick form as incense, in flake form as sachet, and in liquid form as a room freshener and massage oil. The pungent odor of aromatic cedar is generated as oils evaporate from the heartwood. The debate rages as to whether the wood truly acts as a moth deterrent. There's evidence that the wood will kill small moth larvae, but there's more evidence that once the oils evaporate, the wood loses its repellant qualities. A light sanding can renew the odor, but it seems that at least part of the effectiveness of cedar-lined closets and chests is that they're often built very tightly to keep *in* the odor—which in turn keeps *out* the moths.

A few exotic woods have exotic odors. Coachwood has the nose of newly cut hay, while camphorwood smells, not so surprisingly, like camphor. Sandalwood has a spicy odor; those harvesting it will often chop it down, let it lie for several months so ants can eat the inodorous sapwood, then process the aromatic heartwood into various products. Not all is sweet smelling in the world of wood. When the question "What's the worst-smelling wood you've ever worked?" was posted on one woodworking Web site, the opinions were as strong as the odors of the woods discussed.[19] Many thought that first place should go to acacia, with comments like "It smelled like every animal in the neighborhood had taken a dump in my workshop." Other votes went to hoop pine, which "smelt like the worst pair of rotten socks I had ever smelt," and silky oak, which "smells like gone cheese." Several trees, including laurel and *Ocotea bullata,* have earned the nickname "stinkwood." Dahoma wood smells like ammonia. Yellow stercula appropriately derives its name from the Greek word "manure." The agreeableness of satinwood's odor depends somewhat on your ornithological fervor. Albert Constantine explains: "A peculiarity of the wood is that while it burns in an open fireplace very well and with a fragrant restful odor, inducing slumber in many who sit before it, the smoke of this satinwood will kill canaries."[20]

Wood can hurt more than canaries. *Wood: Identification and Use* lists no fewer than 170 woods deemed toxic.[21] Hang around too much cocobolo dust and you might find yourself with conjunctivitis, bronchial asthma, and nausea. Teak dust can cause swelling of the scrotum and oversensitivity to light. If you're having a bad day, try working with American mahogany, black cherry, and iroko; the sawdust can cause giddiness. Dust from the jacareuba tree can cause loss of appetite, and dust from the milky mangrove can cause temporary blindness. White cypress sawdust can lead to nasal cancer and swelling of the eyelids. There's even evidence linking the wood dust from commonplace oak and beech to cancer of the upper respiratory tract. If you're allergic to aspirin, avoid contact with willow and birch; they could cause similar adverse reactions. And, while you're at it, try to avoid splinters from mulga wood, found in Australia; the wood contains a poison that aboriginals use on spearheads.

It is all these different factors—species, density, color, odor, figure—that create the tremendous diversity in wood. Like the snow and snowflakes with which this chapter began, each and every piece of wood is unique.[22]

BAMBOO: THE GRASS THAT THINKS IT'S A WOOD

When workers recently built the seventy-eight-story Central Plaza office tower in Hong Kong, they used what to most Westerners would seem to be a strange material for scaffolding: grass. It was no ordinary "mow your lawn on Saturday" grass but one that grows 100 feet high, is as big

Bamboo scaffolding being used to construct a skyscraper in Hong Kong.

around as a Folger's coffee can, and is harder than oak. It's a grass that doesn't really act, look, or feel like a grass. The grass they used was bamboo—over a hundred thousand poles of it. This was no isolated incident; 90 percent of the scaffold used in Hong Kong is bamboo. And no wonder—according to contractors and builders it costs 95 percent less than steel scaffolding and goes up six times as fast.

Classifying bamboo as a grass (which it is) instead of a wood is akin to classifying a tomato as a fruit (which it is) instead of a vegetable. In both cases, the rightful classifications defy the way each looks, is used, and is discussed in day-to-day language. But that's the way the taxonomic dice roll.

Why isn't bamboo a wood? It's woody, and it's used to make furniture, flooring, bridges, musical instruments, houses, and other wood-like things. Still, it's a grass. Bamboo lacks the cambium layer, the part of the tree responsible for generating the xylem and the phloem, which in the end create official "wood." Bamboo plants don't have trunks; they have clums. A bamboo clum normally reaches full size in a single growing season; after that, it will grow more side branches, and it will grow harder and woodier through a process called lignification, but it will grow neither broader nor taller.

There are over a thousand species of bamboo, and the plant is native to all continents except Europe. It can grow in areas where it's buried beneath snow for months as well as in the tropics. Some species live for a hundred years. There is no faster growing plant than bamboo; numerous types can grow 12 inches per day, and some species have been documented as growing over 40 inches in a single day. A person who is patient can literally watch it grow.

Bamboo, used in Asia for over six thousand years, has had an illustrious career. The carbon filaments that Edison used in creating incandescent light were made from bamboo. It can be used to make rayon, paper, and hundreds of other things.

As a building material it can be used for more than scaffolding. It can be split lengthwise and used in nested fashion for shingles. It can be used for rebar in concrete and for pipe in plumbing and irrigation systems. The latest rage is bamboo flooring, harvested from the giant timber bamboos. On the Flooring Hardness Scale, which is based on how much pressure it takes to embed a 7/16-inch ball halfway into the surface, bamboo rates higher than walnut, teak, cherry, or red oak.

There is no wood more dimensionally stable than bamboo. Its structural flexibility makes it antiseismic; in 1991, the only homes surviving an earthquake in Costa Rica were those of bamboo. If you're into larger-scale examples of bamboo's durability, take a look at the Anlan suspension bridge in China; from AD 300 until 1975, the 1,000-foot span bridge was suspended from cables made of shredded and twisted bamboo.[23]

So damn the definitions, botanists, and copy editors. In my book (literally) bamboo is a wood, and from time to time it will be judiciously spoken of as such.

RESCUING REDWOOD THE HARD WAY

"Pssst, buddy; I wanna show you something" seems to be a strange thing to hear standing in a Midwest lumber salvage yard. But I follow the sawdust-clad stranger down the canyons of wood planks piled ceiling high, chest-wide boards, and Valhalla-size beams sawn from the rescued frameworks of old Montgomery Ward buildings. And in a far, wood-chip-strewn corner of the warehouse, he reveals his secret. It doesn't come by way of a flashing trench coat or a rolled-up sleeve revealing counterfeit Rolexes. It comes by way of a half-dozen 3-inch-thick curly redwood slabs, each 40 inches wide and nearly 20 feet long.[24]

"Even the President of the United States can't get ahold of first-growth redwood like this," he whispers, patting one of the $4,000 slabs. "Not no one." So how did he? Not easily.

When I make a call to Jeff McMullin of McMullin Sawmill in northern California—the man responsible for producing and shipping the redwood slabs—he remembers every detail of those very boards, even though he milled them four years ago. "Oh, yeah, there were seven of 'em, 19 feet long, right? About 3 inches thick and 3 1/2 feet wide, right?" Yes, right. He explains that these monsters came from a butt cut—the lowest section of a tree trunk. But it wasn't even a whole butt; these gems were cut from a mere quarter section of a log. "It was a long triangular-shaped piece of curly redwood we found in the brush behind an old sawmill," he explains. "Back when they first cut redwood they wanted structural wood, and that piece with its wavy grain just wasn't what they wanted." One generation's trash is another's treasure, and this curly redwood—once slabbed, planed, sanded, finished, and polished—is a treasure indeed. Rich ribbons of color ripple across the grain. It has depth and surprise. For some inexplicable reason, it makes you want to take a bite out of it.

Redwood is indeed a wood worthy of all the hooting and hollering. Structurally it's not as strong as more commonly used woods like Douglas fir and yellow pine, but it is more rot resistant because of the tannins and other natural preservatives that congregate at the heart of the tree to strengthen it and add to its longevity as it ages. It's dimensionally stable, little prone to cracking or checking, even when used in exterior applications. It is lightweight, cuts easily, and is well behaved when a screw or nail is driven through it. And though it turns gray when left exposed to the elements, it's a sophisticated, well-heeled silver gray.

So sought after was redwood by bridge builders, water tower erectors, vintners, coopers, and warehouse constructors that early loggers reduced the 2 million acres of virgin old-growth redwood forests by 96 percent. Science writer Richard Preston mourns their demise, "The remaining scraps of the primeval redwood-forest canopy are like three or four fragments of a rose window in a cathedral, and the rest of the window has been smashed and swept away."[25]

Jeff McMullin is a redwood tree prospector. He salvages his redwood where he can find it, and he rarely finds it in convenient places or form. He salvages logs that have been undercut by flooding, washed downriver, floated out to sea, and then washed up on beaches. He finds logs lying where they were left decades ago by loggers who were "high grading"—taking the choicest cuts—and leaving the rest to rot. He salvages redwood timbers from bridges and buildings being dismantled. And he ferrets out standing snags: partially burned or rotted stumps with usable wood. Usable for those willing to work getting to it. He talks about the challenge in felling remnant redwood stumps that are nearly as wide as they are tall. "With a real tall tree, you slant the cut a degree or two and gravity will bring it over. But with a short stump, you cut it through and it just stands there. You've gotta make it tilt a long way before it'll fall over." McMullin gets his tilt using 100-ton hydraulic jacks.

He talks about a redwood log 35 feet long and 9 feet in diameter that he encountered several years ago. He purchased it from someone who'd found it washed up on a sandbar after a flood. After he'd paid for it, the owners of the land it had been uprooted from upriver came and claimed it as theirs. A legal battle ensued, landing the case in court—not once but twice. It became a question of who really owns a tree: the owners of the land where the tree grew or the owners of the land where the tree washed up? Provenance comes into play, even

Redwood snag 14 feet in diameter felled by sawyer from McMullin Redwood. Stumps this wide need to be coaxed into a prone position with hydraulic jacks.

with trees. Especially valuable trees. McMullin wound up with the log, but he wound up paying for it twice.

Cutting logs with redwood-size girth poses logistical problems. For some slabs he uses a pole mill, a horizontal chainsaw with an 8-foot bar with a nose that pivots around a rock-solid center pole. For trees larger yet he uses a humungous Alaskan chainsaw mill, a device similar to a cheese slicer, but instead of a roller guide following the edge of the cheese block while the wire cuts, these rollers follow a flattened part of the log while a chainsaw cuts an even slab.

Even McMullin's largest mill couldn't begin to dissect the largest of the largest legendary redwoods. *The 2006 Guinness Book of World Records* has christened the Stratosphere Giant, at 372 feet, the tallest living tree of any type anywhere. Located in the Humboldt Redwoods State Park, the exact position of this coast redwood is kept a secret to prevent it from meeting the same fate as its predecessor, because being the world's tallest tree is like being Princess Diana—the paparazzi can kill you. In 1963 the National Geographic Society went searching for the world's tallest tree and christened a

367-foot redwood in California the grand champion, naming it Tall Tree. Though its name lacked imagination, the huge tree garnered huge publicity and even huger crowds. In California, in the 1960s, in a redwood forest, countless groups made pilgrimages to the site and felt irresistibly drawn to link hands to surround the tree in a variety of cosmic ceremonies. So many tree enthusiasts circled, paid homage to, and caressed the tree that the ground surrounding it became hard packed. The roots lost access to the water and oxygen they needed. Within ten years the top 15 feet of the tree died, and thirty years after its anointment, the top 10 feet of the tree fell away, forcing Tall Tree to surrender its crown to the Stratosphere Giant.

Redwoods are miniature ecosystems in and of themselves. In the upper crowns of larger trees one can find entire huckleberry thickets thriving. Currant, elderberry, and salmonberry bushes have been found, and to underline the complexity of the redwood environment, different species of *trees* have been found growing in the crowns, including hemlocks, firs, oaks, and, in one instance, an 8-foot-tall Sitka spruce. The largest of the largest contain neatly half a million board feet of lumber and can annually add a ton of wood to their already substantial mass.[26]

McMullin's hard-won redwood lumber winds up in pergolas, trellises, timber frame houses, and gates. It's milled into clear heart bevel siding, jumbo shakes, and lots of natural-edge, Nakashima-inspired dining room tables.

Just how valuable is redwood these days? McMullin talks about a 15-foot-diameter stump he bought for $2,000. By the time he was done cutting sixty slabs out of it, he had almost $100,000 of drop-dead-gorgeous curly redwood: a return on investment even Warren Buffet could appreciate.

LOGGING THE INDUSTRIAL FOREST

The Duluth Timber Company, where I was first introduced to McMullin's slabs, dabbles in redwood, but its stock in trade is reclaiming lumber from yesteryear's buildings.[27] Much of the timber the company salvages is from warehouses, shipyards, and munitions plants built when old-growth pine and fir dominated the land. At the Minnesota sawmill they've handled beams as large as 50 feet long, 16 inches wide, and 50 inches deep.

They call it logging the industrial forest. Their literature explains: "This is the forest we are harvesting today. Posts and beams with the length and dimensions of the old-growth trees; prime timber, tall, sound, structural. Joists and planks trod by millions of workers, straight grained lumber from tanks that held water, wine, and vinegar." For the most part, old buildings on the East Coast and central United States yield white pine, while those on the West Coast yield fir. The exceptions are buildings built during the World Wars, when metal was so scarce that everything—East Coast, West Coast, and everything in between—was built from structurally superior fir.

"When I started here ten years ago, demolition companies just gave us the wood, happy to get it off their property. Or maybe sold it to us for a few pennies a board foot," explains sales and marketing manager Liz Bieter. "Now we sometimes have to pay up to $2.50 a board foot." The thirty-some companies now in the reclaimed timber business have awakened the law of supply and demand. Even with flooring and timbers selling for as much as $10 a board foot, it's not an easy business.

"We're always chasing the wood," explains Liz. And after they catch it, there's plenty of work to do. Liz and I walk through acres of timbers and joists stacked outside the warehouse, from time to time catching a glimpse of a taconite-hauling behemoth plowing its way through Lake Superior. We come across a pile of 12 × 12 nail-infested beams stacked next to a massive bandsaw with a 36-foot-long outfeed table. Before the beams meet the bandsaw, each beam will be denailed, scanned with a metal detector, then denailed again. It's not only nails that get denailed; wire, knife tips, and bullets have also been found. "But metal detectors don't pick up rocks and concrete, and those might as well be metal when they hit a blade," explains Liz. Some timbers from some munitions plants must be bypassed because they contain such a high concentration of gunpowder they could explode if overheated while being cut.

Reclaimed lumber costs about a third more than newly milled lumber, yet the demand remains strong. Why? For starters, if you need truly massive timbers, there are few harvestable trees of that size remaining. Most buyers like the romance and appearance of reclaimed lumber. There are others who are environmentally committed and like the idea of cutting existing timbers to size instead of cutting down live trees. Because of the expense, many people use reclaimed wood for accents: a mantle, staircase, or exposed beam or two.

Bill Gates and Microsoft have spawned thousands of satellite companies, but none as strange as the Duluth Timber Company branch in Edison, Washington. While supplying reclaimed timbers for Gates's house in Washington state, the Duluth Timber Company found itself dismantling buildings on the West Coast, shipping the timbers to Minnesota to be resawn and milled, then shipping them back again. They decided to open a West Coast branch when the project began calling for timbers too long to fit on a truck. And just how big is big when it comes to Bill Gates's beams? They're 16 inches wide, 55 inches tall, and 70 feet long—a total of 5,005 board feet per beam. In comparison, the average *house* built in the United States contains just over 13,000 board feet of framing lumber.

Beyond gunpowder-filled beams, there are other hazards of the business. The Duluth Timber Company searched high and low for a local sawyer to mill flooring but came up empty. No one wanted to risk their equipment—particularly the expensive saw blades—cutting up nail-infested lumber. "Finally we found a guy who'd do it, but he told us right up front, he'd charge us $200 every time he hit a nail," explains Bieter. "And, you know, it's almost guaranteed if there's a nail anywhere in a timber, you'll hit it." Though they attempted to ferret out all the nails before sending out the timbers, at the start the sawyer would walk in every few weeks with a handful of nails, and walk out with a large check. The company finally hired a guy who was the Fred Astaire of pulling nails. He was so good the sawyer didn't hit a nail for twelve months. He was the consummate profes-

A gigantic, complex scarf joint being used to join two massive timbers at Bensonwood—a timber framing company.

sional. At one point he moved to Alaska, but he moved back because he missed pulling nails.

In addition to the Duluth Timber's brand of wood salvage, there are other types. There's forest salvage, like the type McMullin executes. There's waste-stream salvage, a labor-intensive process that involves rescuing lumber from demolished houses. There are salvage operations that specialize in dismantling barns, bridges, and railroad lines. There are others that dredge up sunken logs from the bottoms of rivers and lakes. If you add them all up, you have about sixty companies cranking out 250 million board feet of recycled lumber annually: an impressive number, but one that still accounts for less than 1 percent of the lumber sold yearly in the United States.[28] But Duluth Timber has found its niche. And the million board feet of timber piled outside the door—just waiting to be reborn as mantels, flooring, and beams for the next Bill Gates house—attest they'll be in that niche for a long time to come.

WOOD: HOW IT GOT HERE, HOW TREES MAKE IT

If you were to grab your camera, step back 3.8 billion years, climb a steep hill, and look through the viewfinder, you'd see a lot of rock, volcanic activity, open space, and water. You'd also discover you were panting like a St. Bernard because of the oxygen-starved environment. But more conspicuous than what you would see would be what you would NOT see: no plants, no shade, no greenery. The earth would be, well, all earth tones.

But if you were to walk to the edge of the water and slip on your close-up lens, you would see something heart-warming: the first single-cell life-forms. Not very big, not doing much, but hey: a start. The exact origin of these miraculous cells we leave for Darwin, Jimmy Swaggart, and the Discovery Channel to debate, but the undeniable fact is that they're there. Click the shutter and voilà: you've got a photo of where the story of wood begins.

You'd need to wait around another five hundred million years or so before you'd start seeing blue-green algae form. These were also single-cell entities, with the BIG difference that they'd learned a seemingly magical new trick: they'd found a way to harness the sun's energy to produce food. They'd learned how to photosynthesize.

Eventually some of the algae evolved into thin platelike organisms. Their shape made them excellent at floating and photosynthesizing. They flourished and kicked out lots of oxygen, but as smart and industrious as these plants were, it still took them another 2.5 billion years to figure out the land thing. Part of the reason for the delay was that living in water was living the life of Riley. The ocean and river currents mixed oxygen, water, and minerals into a primordial energy drink that was easy for plants to access. Moving to land required pulling oxygen out of the air and extracting water and minerals out of the soil. There was a great risk of dehydration. And then there was this thing called gravity.

The movement to land was probably spurred by competition for sunlight. At some point a group of cells—most likely in their adolescent years—rebelled against the status quo and decided, literally, to stand up for themselves. Some algae began issuing rootlike structures that crept toward dryer land, eventually producing small shoots with pinheads. These shoots stayed close to shore where water was plentiful, and they began the long hard struggle against gravity in order to rise up and out of the water. The first upward-reaching shoots supported themselves through turgor—the process whereby the outer shell of a plant becomes stiffer by being filled with fluid. Or, as one of these adolescent plants might have snickered, "I think I'm getting a woody." Which was, as it turned out to be, a much more prescient and astute observation than he'd ever realize.

The earliest known land plant, *Cooksonia,* evolved around 450 million years ago. They grew only a few centimeters tall but were probably the first land plant to have a vascular system. Next, *Rhynia* plants developed a sturdier stem by joining rows of cells end to end to form pipes that could supply water to the rest of the plant. This early vascular system was a key player in the evolution of trees. It allowed plants to grow taller. The vascular system began incorporating a chemical called lignin, a glue of sorts, which made stems even more rigid. Roots grew longer and stronger to supply water to the vascular system and primitive leaves. Now there were roots and a sturdy woodlike stem, all topped off with photosynthesizing greenery. BAM, all of a sudden (well, not really) you had an organism capable of creating a substance remarkably similar to (but not quite) wood.[29]

Around 370 million years ago, three types of trees began evolving in earnest. As a group, they're often referred to as the coal swamp trees—a name vaguely reminiscent of one of my son's early heavy metal bands. They all looked tree-ish, but each had a different root, trunk, and leaf system—and none produced "wood" per se, at least as we know it today.[30]

Club mosses developed scalelike leaves to improve photosynthetic efficiency and thicker stems for improving water distribution. These stems, containing water-conducting xylem and food-conducting phloem, were strengthened by a barklike material that encased them and grew as the plant grew. These were the first plants to grow permanent trunks.

Horsetails developed stems that were more like hollow tubes, surrounded by a xylem ring, then an outer phloem ring. Regularly spaced bulkheads divided the stems into compartments, which gave them a look similar to bamboo.

Ferns, the third type of early tree, excelled in the leaf area but didn't pay much attention to developing good trunks. They could grow vertically only by supporting themselves with tangled aerial roots and leaves. These looked, and still look, a lot like palm trees. No ninnies, these grew to a height of 60 feet or more.

In a world with plenty of sunlight and water, and not much to stand in their way, these trees reproduced and grew unabated. The math shows that as many as ten million generations of behemoth trees could have come and gone in the three hundred million years from the time these trees first evolved. The early coal swamp trees gained their nickname because they were kick-ass at making coal. They lived fast and died hard (yuk, yuk), keeled over into the oxygen-poor swamps, and partly decayed into layer upon layer of peat, which eventually, under vast geological pressures, turned to coal.

Two other groups of trees eventually decided to branch out. They were more adaptable and didn't use the touch-and-go method of reproduction used by the coal swamp trees, which relied on the dispersion of spores in water. These new trees developed seeds—seeds that could stay attached to, and be nourished by, the parent tree before dropping to the ground or being whisked away by wind or birds. Dr. Phil might say these seeds "had trouble letting go," but this new development went a long way in helping trees evolve.

Botanists today divide these trees into two large groups based on the types of seeds they produce and how they reproduce: gymnosperms, which means "naked seed," and angiosperms, which means "enclosed seed." Woodworkers divide them into two groups, too: softwoods and hardwoods, respectively. They do this even though some softwoods are harder than hardwoods. Softwoods like southern yellow pine and Douglas fir are much harder than hardwoods like basswood

Conifers evolved around two hundred million years ago and today lay claim to being the heaviest, tallest, and oldest living things on the planet. This well-protected conifer—a Wollemi pine at the Royal Botanical Garden in Sydney—was discovered in 1994 in a remote valley of Australia and can also lay claim to being among the rarest.

or cottonwood. And balsa—a hardwood? Give me a break. Balsa is so light it's used to build gliders.

Gymnosperms evolved first—around two hundred million years ago—and were able to change and adapt to the geographic and climatic swings that have taken place right up through today. It's the descendants of these cone-bearing trees that hold the Academy Awards for oldest, tallest, and heaviest living things. Thank God they're the tall silent type, because their acceptance speech would go on and on. "I'd like to thank my mother, my grandmother, my great-grandmother, my . . . " and on back for the millions of years they've been in existence.

Angiosperms—flowering trees and plants—evolved later than the conifers; around 125 million years ago. But it was worth the time they spent in the R & D department: 275,000 species evolved, nearly 80,000 of them trees.[31] They also made structural advances. Conifers and other gymnosperms developed only one type of cell for transporting water to the leaves, and that cell had to serve double duty: it was responsible for transporting water to the leaves and for reinforcing the trunk. The angiosperms, however, developed specialized cells, some for transporting water, others for strength. Put these advances together and you've got one Ferrari of a tree. And angiosperms *were* versatile and fast. They evolved into the oak, beech, elm, chestnut, and other trees that produce the wood woodworkers love to cut up.

Angiosperms and gymnosperms coexisted peacefully side by side for millions of years. And why not? There was plenty of water and sunlight. There were few predators. And it was warm. Even though—from time to time—it did get a bit chilly.

So there you sit, an honest, hardworking planet. You've done your best to be a good creator and provider. You've waited patiently—billions of years, actually—for trees to form. You've endured volcanic eruptions, colliding tectonic plates, and hot flashes so intense you could fry a Hypselosaurus egg on them. You've raised ten million generations of trees—some not all that well behaved—and cleaned up more dinosaur shit than you can shake a frond at. You've planned, sweated, and sacrificed in order to create this thing called wood.

But then, uninvited, these fresh young things—these ice ages—just saunter in and proceed to wipe out many of the trees and forests it's taken you millions of years to create. You can almost hear them say, as they slide on the white gloves, "Okay, old world, you've taken care of the dirty work; now we're going to do a little interior decorating," then turn to one another, wink, and say, "A little something to freshen the palate?" Less and less of the earth remained habitable. An asteroid or two that struck the earth with the force of a few million megatons, blowing billions of cubic feet of dust into the air blocking out the sun, surely didn't help matters. Soon it was good-bye Golden Age of Trees; hello Permo-Carboniferous Ice Age.

These ancient ice ages had a great impact on much of the earth's flora and fauna, but a wave of four more ice ages that began descending a million and a half years ago played a disproportionate role in shaping the world's forests as we find them today. The glaciers came and went in long, cold spurts interspersed by short warming periods lasting sixty thousand years or so. All of this involved a lot of pushing, shoving, freezing, melting, and compacting.

The last glaciers, which receded about fifteen thousand years ago, changed the landscape considerably by the routes they traveled and the impact of their enormous weight. As they moved and melted they established the size, depth, and location of many lakes, including the Great Lakes. The ground and rock they plowed before them established the depth and richness of the soil and, in turn, the plants that would thrive or die. They even pushed around enough glacial till to form Long Island.

In North America, where the mountain ranges were aligned primarily north to south, many trees were able to survive by reestablishing themselves as they were "running down the valleys" ahead of the encroaching glaciers. And they were able to easily recolonize in the wake of the receding glaciers in the rich soils left behind in the valleys. In the United States today, 1,182 different species of trees can be found.[32]

Trees in Eurasia encountered more roadblocks. They were forced to climb "up and over" the east-to-west-oriented mountain ranges: a much more difficult task. Fewer survived. Thus, one theory submits, the more limited number of tree species in Europe and temperate Asia than in North America and some other parts of the globe.

Ten thousand years ago, glaciers had retreated so far toward the poles that forests once again had solid toeholds in most parts of the earth. Some areas that were once swampy were now desert. Some areas that were bone dry now had water. In a few short years—at least in terms of the earth's life span—the entire greenery of the globe had changed. It was a fresh start.

This whole discourse on trees may seem to be quite a circuitous route to the lumberyard, but the point is this: Wood was a long time coming, and it's seemingly here to stay. If you look at a globe that's been color-coded according to tree types and zones, it looks like a swirly marble. Yet, on closer examination you can see some general patterns in terms of trees and the woods they create.

There are *rain forest* trees—mostly hardwoods, like ipe and mahogany—that grow continuously and are so "ever green" that few produce nice consistent growth rings like those in areas where seasons and temperatures fluctuate more. Some tropical trees in ideal conditions can grow up to 15 feet in a single year. Tropical trees often have narrow trunks (less wind means less need for thick trunks), small crowns (there's plenty of sunshine and not as much tree mass to feed), and thin bark (no need to protect against frost damage).

Trees of the *seasonal* or *monsoon* forests are subject to extreme wet-dry cycles. These climatic extremes produce trees with some of the densest, hardest, and most beautiful woods around: ebony, rosewood, and others.

Savanna and *desert* trees live in areas that receive little rain. Trees still grow, but in limited numbers. Some developed bizarre survival mechanisms, resulting in bizarre appearances; you'll find acacia trees with roots 160 feet deep and saguaro cacti with trunks that can swell to store 250 gallons—one ton—of water. These bizarre survival mechanisms create some pretty eccentric woods.

The *temperate* forests contain trees that many people would call normal. These are the forests that dominate much of Europe, North America, and northern Asia: oaks, maples, and pine. Trees can be deciduous or evergreen, and each has its own survival strategies. Deciduous trees are more active; they grow wide leaves that frolic hard in the exciting world of photosynthesis during the warm months but drop them in order to limit water loss during the winter. Evergreens are more steady-as-she-goes, with needles that are less adept at photosynthesis but remain intact year round, since they are tougher and more drought tolerant.

In *boreal* forests closer to the poles, gymnosperms are the dominant tree, primarily because of their superior frost- and drought-tolerant needles. Clever shape, too; their conical shape allows their downward-sloping branches to slough off heavy snows as they accumulate.[33]

The important thing to take away is this: Trees have adopted different survival and growth mechanisms throughout time to cope with their environment. Those adaptations have manifested themselves in the roots, flowers, leaves, and, most important for us, branches and trunks, that create the amazing variety of woods we have today. Many trees needed to develop ways to defend their wood against attacks by fungi, and thus they developed chemical preservatives that create the rich, dark lumber woodworkers love. Boreal trees, like pine and fir, in their competitive race for sunlight developed the strong, arrow-straight trunks used by carpenters to build houses. And unique adaptations like this produce the thousands of unique woods, which have been crafted in millions of ways, by billions of people for hundreds of thousand years. Wood guru Bruce Hoadley reminds us in *Understanding Wood: A Craftsman's Guide to Wood Technology*, "Wood evolved as a functional tissue of plants and not as a material designed to satisfy the needs of woodworkers."[34] And though this is true, aren't you glad trees evolved so they produce wood, rather than something like, say, Silly Putty?

But what exactly is wood, anyway?

A tree is an improbable thing. Its overall architectural and biological design would seem to doom it from the start. A narrow trunk topped by a gigantic sail made of leaves would seem to make it a structure destined for toppling in even a modest wind. The distance between the roots, which gather the water, and the leaves, which transform that water via photosynthesis, is a long, gravity-defying journey—in some cases a trip exceeding the length of a football field. Their initial growth, a glacial pace compared with the shrubs and weeds around them, would seem to set up a battle for sunlight and water they could never win. Yet, trees have not only survived but thrived. There are tens of thousands of species, ranging in size from pint-sized to over 300 feet tall. Trees inhabit 3 of every 10 square miles of dry land. It could be said with a good degree of certitude that no other element has had such an intense and profound effect on the planet and its inhabitants as trees—both as a natural phenomenon and as a raw product for making damn near everything. Since it's wood we're interested in here, let's look at the trunk—the part that constitutes 60 to 70 percent of the mass of most trees.

As we've seen, plants originally invented the entire idea of wood as a way of strengthening their stems to rise above the surrounding vegetation and capture more sunlight. So what is wood? Wood is composed of a great number of hollow cells (1 cubic inch of wood can contain four million of them) made mostly of cellulose and glued together with lignin. The cells that function as vertical pipelines for transporting water are arranged end-to-end and side-to-side. They're long narrow cells—tracheids, or vessels—that, depending on the tree, can range from microscopic to those that can be seen by the naked eye. This network of cells not only strengthens the tree but does two other very important things. Some of these cells, constituting the xylem, carry water from the roots to the leaves and other parts of the tree. Other cells, constituting the phloem, carry sap, or food, that the leaves have created back to the rest of the tree. But where do these cells come from? From a thin layer of actively growing and multiplying cells between the bark and the wood. This layer, the cambium, is the slippery stuff you find when you peel the bark off a tree in the spring. It produces wood, or xylem cells, to the inside and bark, or phloem cells, to the outside.[35]

Anyone on the street knows you can tell the age of most pine trees by counting the annular rings on the end of a log. Rings form because the xylem's cells are thin-walled and wider in the spring while growth is quick (the light-colored ring) and narrower but thicker-walled later in the season when growth slows (the darker ring). Except for the most recently growing outermost group of cells, these annular rings are composed primarily of dead cells, though they continue to support the tree and some continue to conduct sap.

The cambium produces phloem cells to the outside. Unlike the xylem, which uses older dead tracheids to help conduct business (and water), the phloem is a one-man show. The active working phloem is close to the cambium; the dead phloem is pushed outward and—along with other tissues and cells—forms the bark.

For the most part, wood is a tight-knit community of cells made of two basic substances: cellulose and lignin. The cellulose makes wood flexible, and the lignin (the glue that binds the cells together) makes it stiff. Together, this dynamic duo gives wood a remarkable combination of qualities: strength and flexibility. Trees have differing amounts of lignin and differently sized wood fibers that dictate their strength, hardness, and flexibility. For instance, the yew tree, with long fibers and little lignin, is flexible and strong, perfect for crafting bows, while the black locust tree has the opposite qualities—short fibers and lots of lignin—and has wood so stiff it can be used as a hammer. But it's these differing qualities that make wood so versatile and so fascinating that someone should write a book about it.[36]

Tree trunks and the woods they produce are strong and tough because they incorporate many of the same principles that make plywood and fiberglass strong. The annular rings, in a general sense, are laminated plies of wood that bind themselves to one another year after year. On a more microscopic level, the millions of long continuous cellulose fibers embedded in lignin serve as nature's version of a composite material, sort of like fiberglass. These fibers don't only hold up trees; they hold up our houses, chairs, and everything else made of wood as well. In fact, the book you're now reading is made up primarily of wood fibers.

As trees age, the older cells at the core of the trunk lose some of their ability to conduct water. The tree allows these innermost cells to retire. These heartwood cells often become filled with resins and gums that help repel fungi and insects, creating the naturally rot-resistant heartwood. This stiffened heartwood core also continues to help structurally support the tree. The heartwood often has a richer, deeper color—a factor of great aesthetic importance to woodworkers. Here a tree honors its elderly cells by letting them rest but still giving them something meaningful to do. We non-trees could take a lesson from that.

Bark is essentially body armor for the fragile active phloem and for the tree as a whole. It protects the tree from insects, herbivores, fire, and rapid temperature swings. Some bark is so protective and ingenious that it produces its own protective mechanisms.[37] Rubber trees exude latex to clog up the mouths of predators. The bark of the quinine tree produces a chemical that's poisonous to some attacking insects (but beneficial to man when used to treat malaria). Pine trees exude a sap that can literally encase attacking bugs.

One particular tree trunk mystery involves how water can make the gravity-defying journey from roots to leaves—in extreme cases a distance of 300 feet or more. The mystery becomes even more baffling when one realizes that a large tree may need to pull over 100 gallons of water a day up its trunk.

There have been various theories. One was that roots "pumped" water up the tree, a theory discredited, since trees don't spout water from their stumps after being cut down. Another theory was that water was wicked up the tree, much the way water travels up a towel when its end is left lying in a puddle. But this type of capillary action can pull water only a short distance. Some posited that water was drawn up the trunk through suction, like milk being drawn up a straw. The problem here is that there's no one and no thing to do the sucking.

However, this last theory comes closest to the truth, for while water isn't sucked up the tree, it is literally pulled up the tree. When the minute pores, or stomates, in leaves are open, water evaporates from the leaf. Each water molecule is part of a continuous unbroken column of water traveling through vessels, or tracheids, reaching all the way to the roots. The water molecules adhere to one another through cohesion; each time a water molecule evaporates from a leaf, it pulls the entire chain of water molecules up behind it. At the roots, a molecule of water is pulled out of the soil to fill the gap in the chain. In other words, the roots, trunk, and leaves form a continuous system that connects the soil below to the air above.

And here's the bottom line. Trees are supremely efficient, self-contained wood manufacturing plants. They're carbon-powered perpetual motion machines that absorb carbon dioxide, convert it to wood, store it for decades, and then release it back into the atmosphere as they decompose or burn, starting the cycle all over again. They're water wheels with roots that extract moisture from the ground, trunks that transport it, and leaves that release it back into the atmosphere, where it eventually falls back to earth in the form of rain, starting the process all over again. It's a full-time job being a tree.

Trees are an in-your-face example of nature's engineering genius. Each part contributes to, and benefits from, the whole. They take three essentially invisible items—carbon dioxide gas, water, and sunlight—and craft it into the solid wood that forms your house, pencil, and baseball bat. That's pretty amazing. As my friend Cliff once said, "It's hard to improve on a tree; Mother Nature knew what she was doing."

So next time you hack into a piece of wood—whether it's a forty-year-old piece of birch, a fifty-thousand-year-old slab of kauri wood, or a two-hundred-year-old Douglas fir timber—pay a little respect.[38]

Running your palm across a two-hundred-year-old tree is cause for reminiscing; to think about how life has changed in the two centuries since that tree began life as a seedling. Back in 1800 when it first took root, the population of the United States was just over 5.3 million; less than 2 percent of where it stands now. And a furniture maker by the name of Duncan Phyfe was just beginning to make a name for himself in New York City. As long as there have been wood and tools to work that wood, there have been woodworkers. Here are a few of their modern-day stories.

CHAPTER 2

The Wacky World of Woodworkers

Woodworking is big. Really big. The ranks of amateur woodworkers in the United States alone constitute an army seven to ten million strong, annually spending over $1 billion. Statistically, it's a guy thing; on the average, 95 percent of those who subscribe to woodworking magazines have a Y chromosome. And it's a middle-aged thing; the average age of those Y chromosomes is 56 years old.

But an average is just that—an average. You'll find eighty-five-year-old woodworkers like Paul Smith of Florida, who specializes in making dominoes, and three-and-a-half-year-old woodturners who work comfortably on a lathe. You'll find specialty groups like Women in Woodworking, Woodworking for the Blind, the International Association of Penturners, the Scrollsaw Association of the World, Wheelchair Woodturners and Woodworkers, and individual woodworkers of every race, creed, height, weight, number of digits, occupation, skill level, and income level.

Woodworking is big on the professional level as well. While overseas outsourcing and computer-guided machines have put some professionals out of business and turned others more into woodworking machine programmers than craftsmen, you'll still find professional woodworkers plying their trades in hundreds of diverse fields.

There are odd pockets galore in both the professional and the amateur worlds. You'll find woodworkers focused on building rocking dinosaurs, wooden locks, "ships in a light bulb," twig furniture, pet furniture, and lathe-turned cowboy hats you can wear. To hear a good woodworking story, all you need to do is clean the sawdust out of your ears.

A CHAINSAW ARTIST A CUT ABOVE THE REST

So, how do you find a chainsaw artist in Minnesota? According to the advice of one seasoned carver, it's easy. "You fly into Minneapolis, rent your car, buy a six-pack of beer, and start driving and drinking. When you have an empty can, throw it out the window

Three-time holder of the World's Champion Chainsaw Sculptor title, Dennis Roghair poses with his grandson, Jake. The two have significantly different carving styles (and hairstyles).

and you'll hit a chainsaw carver." But you won't hit one the caliber of Dennis Roghair. Over the past two decades, Roghair has held the title of World's Champion Chainsaw Sculptor three times and has won awards as varied and strange as World's Best Grizzly Bear, World's Best Bear Carved in Tree, and World's Best Golfing Bear.

He describes one carving that won a competition. "It was a grizzly bear standing on all fours on a sandbar, then a stream and another sandbar where there's a sockeye salmon with an eagle next to it. I named it 'First come.'" It's a remarkable sculpture, made even more remarkable because the entire thing was carved out of a block of wood the size of a carry-on suitcase.

Roghair does a lot of bears. In fact, almost all chainsaw carvers do a lot of bears because that's what sells. Owls, eagles, and elves are right up there, too, but bears pay the rent. But when leaving the land of bears, Roghair has a tremendous repertoire—mammoth, dinky, and everything in between. One of his most dramatic pieces, standing just outside the door of his shop in Hinckley, Minnesota, is a 12-foot-high sculpture of a pouncing mountain lion clawing the haunches of a leaping bighorn sheep. A life-size sculpture of a white tiger sits curled up in the back of his store. For one suburban homeowner he crafted a working lamppost complete with a life-size lamplighter leaning against it, lighting his pipe, with a mini-light inside the pipe that glows.

When he began chainsaw sculpting thirty-two years ago, only thirty to forty people were doing it. Now, according to Roghair, there are "a hundred good sculptors and a zillion really bad ones." One of the qualities that make Roghair one of the good ones is his ruthless pursuit of realism. A few years back he was commissioned to do a life-size sculpture of "Ole" Olson; a long-time ground crew employee of the Minnesota State Fair who'd recently passed away. He worked from old pictures, but somehow, when the sculpture was complete, he felt he hadn't captured the true Ole. "I kept looking and looking at the pictures and realized it looked like he was missing a thumb. So I made some calls and found out he *was* missing a thumb—so I cut his thumb off. Later I got a nice letter from his family telling me how much they appreciated the sculpture right down to the missing thumb." Ruthless pursuit of realism.

The worst woods Roghair has ever carved are eucalyptus in Florida and live oak on the East Coast. "After five minutes your saw is dull and you're working in a cloud of blue smoke." He prefers to use white pine; a wood that's plentiful where he lives and works. He uses freshly cut logs because green wood carves faster and cleaner. As the sculptures dry, sun and wind cause the outer surface to dry and shrink faster than the center. Since Roghair "guarantees all my sculptures will crack," he strategically carves a deep groove in the back of most of his works to relieve the pressure and give the sculpture an inconspicuous place *to* crack.

Roghair turns lots of still-standing damaged and dead trees of sentimental value into custom sculptures. Initially, he carved the trees as they stood, but the root system would continue to wick up water, and they'd rot from the inside out. Now he severs them completely and shims them up just a hair—less than a half inch—so they last. Left outdoors in its natural state, a white pine sculpture has a life span of ten to fifteen years. But properly preserved, it'll last nearly forever. Roghair recommends a preservation system based on three-year cycles: in the first three years, apply a preservative twice a year; for the next three years, apply it once a year; after that, apply a preservative once every three years.

In the world of chainsaw artistry, anything and everything can be, and has been, made. People carve Elvis, Jesus, W. C. Fields, cowboys, "Harley Gothics," hockey goalies, horses, seahorses, herons, gigantic pine bark beetles, baseball bats, dragons, chairs, and huge fish swallowing coyotes. Chester Armstrong's highly polished walnut wildlife sculptures look like Remington bronzes, while R. L. Blair's animals seem to have stepped right out of the pages of a Dell comic book. Glenn Greensides likes to work big, with 16-foot-tall sculptures of grizzly bears, fly fishermen, and hikers, while Wild Mountain Man Ray Murphy carves wristwatch-size ladybugs. Beyond starting the process with the same tool, there aren't too many threads holding the league of chainsaw artists together. Some use chainsaws from start to finish, while others use anything and everything, finishing up with tools as refined as dental picks.

Roghair does most of his work using only a handful of tools. He normally starts out with a chainsaw with a 2-foot bar, though for some of his larger commissions he'll use bars up to 5 feet in length. Once the sculpture is roughed out, he switches to a smaller saw—one with a 12-inch-long sculpting bar and a dime tip. The "dime" refers to the radius of the nose—a nose so small that it requires a special chain with short links and modified teeth to make it around the tight curve. The other tool he often uses is an eyeball tool, a bolt with a head that's been scooped out, which he chucks into the jaws of a high-speed die grinder. The tool burns the wood as it shapes the eyeballs, making the wood harder, smoother, and glowing, so the eyes look lifelike.

Like most people who have mastered their trade, Roghair makes the whole thing look easy. He has the same attitude and approach as Michelangelo, who talked about removing pieces of stone until the figure inside revealed itself. Inside the blocks of marble Michelangelo worked, he found *David,* the *Pieta,* and the *Battle of the Centaurs.* Inside Roghair's blocks of wood, he finds abundant wildlife.

In a wood carving class I took a few years back, I wasn't as lucky. The figure that was supposed to "reveal itself" from within our blocks of balsa was a jaunty-looking sailor with his hands in his pockets and a pipe jutting out from the corner of his mouth. Most of the blocks of wood the *other* students carved did contain a jaunty-looking sailor with a pipe in his mouth. But inside my block of wood there was someone else; the sailor would not reveal himself. As others carved finely detailed sailor hats, tight-fitting britches, and wry expressions on salty faces, I carved baggy overalls. Eventually, as my knife slipped more and more and my frustration grew and grew, my block of wood revealed what was truly inside: a farmer with a load in his pants who'd gotten his arm stuck in a combine, because he had only one. Working wood is a gift few people have; Roghair sits firmly in that group.

* * *

The degree of detail that top-notch carvers can attain using only a standard chainsaw is impressive. They can smooth a surface by laying the bar flat, raising one side slightly, and sweeping the bar back and forth. Fur is created by chopping with the tip held sideways to the surface; feathers by making quick overlapping tip strokes; eye sockets with quick sideways sweeps of the tip

There are surprises to be encountered while carving. Roghair has hit lots of fence wire and ceramic insulators. He talks about one carver who was carving a sailor out of a tree crotch when he encountered a logging chain that the tree had engulfed decades before. With an attitude of turning lemons into lemonade, the carver integrated it into the sculpture as a pocketwatch chain.

It's an art form more dangerous than most. In over thirty years of carving, Roghair has never had an accident, but he's seen them. One carver knocked two front teeth out with a chainsaw that kicked back. Another artist slit his mouth all the way to his ear in the same way. Not that he hasn't done dumb things: one time he carved the name of a TV host into an apple while the host clenched the apple in his mouth.

One legendary chainsaw sawyer, the Wild Mountain Man, puts on shows during which he saws people's names in wood belt buckles while they're wearing them—a stunt he's performed ten thousand times without disembowelment. "It's good to get someone with a firm stomach, calm nerves, and a short name," he explains.[1] He hasn't been so careful in terms of his own life and limb, having encountered over thirty serious accidents, hundreds of stitches, and the loss of two fingers. He's sawed over fifty thousand works of art, some created during a stunt in which he uses two chainsaws to simultaneously carve two sculptures. The Wild Mountain Man can saw out a chair in 10 seconds and has works displayed in every Ripley's Believe It or Not Museum in the world; specifically Number 2 pencils with the entire alphabet sawed into them. Perhaps the only feat topping that is his ability to saw ten numbers on a toothpick. With a chainsaw, mind you.

Wild Mountain Man Ray Murphy peeks at a pencil upon which he's sawn the entire alphabet with a chainsaw. He's also sawn sixteen numbers on a toothpick, and names on belt buckles (while people were wearing them), using his tool of choice.

Roghair's largest project, based on size, length of time, and amount of red tape, is the 26-foot-tall voyageur standing at River Side Park in Pine City, Minnesota. The total length of the redwood log used—part of it buried for stability—is 40 feet, with the figure measuring 8 feet across at the shoulders. He explains it was like carving a silo, and at the start he was using a chainsaw with a 5-foot bar. From start to finish the statue took almost four and a half years. The mammoth redwood log, valued at $25,000, was found, of all places, in a ditch behind a lumberyard in central Minnesota. The project was to be installed in a city park and was being overseen by a committee under the guidance of the historical society. First he was going to carve the statue in one place, then another. "It involved four years of haggling and four weeks of carving," Roghair explains.

As I prepare to leave, his 19-year-old grandson, sporting a modified Mohawk, appears. He too has taken up chainsaw carving, favoring dragons and fantasy creatures over the more earthly sculptures his grandfather and mentor sculpts. He's pretty darn good, too. But Roghair can't help but rib him. "This is my grandson, Jake," he says by way of introduction. Then, turning to him, he says, "Or maybe you've changed your name now to World's Best Chainsaw Carver?" And I think to myself, "No, I believe that title's been taken."

MY SEVEN AWKWARD MINUTES WITH THE MAN WHO CARVES FERRARIS

My son, Zach, and I walk around a block in Venice, Italy, three times looking for the workshop of Livio De Marchi—and in doing so walk right by his store three times. The problem is that Livio is a woodworker, but his store windows are filled with hanging coats, golf bags, hats, and purses; all meticulously crafted of wood, realistic right down to the stitching on the coat and the zipper on the bag.[2]

Inside the store we find cowboy boots, paintbrushes, and suitcases—all wood. The monster books that form the legs of the desk, the wand blowing bubbles, the chest of drawers—not just the chest, but the stomach, groin, and legs, too—all wood. His latest works incorporate glass, and we find gigantic tubes of toothpaste, miniature balloon-borne gondolas, and Technicolor paintbrushes. But mostly it's wood.

Livio steps through the door leading from his workshop, and I shake hands with a man who resembles Salvador Dali, complete with handlebar mustache and artistic flair but lacking the psychosis. His wife joins us, and try as we may, the four of us quickly discover we have no common linguistic ground. We do manage to communicate to a small degree with facial expressions and "point language." By pretending to press the plunger on my camera and pointing to the door behind Livio, I try to get across the notion that I'd love to take a picture of him in his workshop in back. And with all the kindness and exactitude of a man who is going to say "yes," he says "no." Which kills me. Because while the wood sculptures in his storefront are fabulous, his larger works—the things that *must be* in progress in his workshop just feet away—are magnificent, outrageous, unbelieveable.

Livio De Marchi, looking very much at home in one of his *Casa die Libri* (house of books). Everything—chair, wood stove, curtains—is carved of wood.

I first became acquainted with Livio's work via a series of photos that circulated on the Internet—if you're on the "forward" list of any woodworker, you've seen them too. You see Livio seated in a Ferrari F50, made all the more fabulous by the fact that it's exactly replicated in wood, made more fabulous yet that he's driving it, made even more fabulous that he's driving it in the Grand Canal in Venice. It's amphibious, which is the kind of vehicle you want in a town with no streets. It is exact, down to the sun visors and emergency brake lever.

Other projects he's done over the past twenty years are also studies in the seemingly impossible. In 1986, he crafted a floating woman's shoe, commandeered by sixteen stripe-shirted gondoliers. In 1990, he built the first of three *Casa die Libri*—houses of books—constructed of, yes, books of wood. He intends to construct seven more. Inside the various houses you find a stove of wood, complete table settings of wood, a mammoth book of wood that opens into a bed, wood socks hanging from wood clothespins, wood fruit, wood bookshelves holding wood books, and even wood people.

Wood sculptor Livio De Marchi driving his all-wood amphibious Ferrari F50 in the Grand Canal in Venice.

In 1994, he crafted "A Dream in Venezia," which consists of a Cinderella-size pumpkin coach with four horses that rear when the reins are pulled—motor driven and aquatic. Over the years he's crafted an amphibious Mercedes Seagull, a Volkswagen Beetle, a Jaguar, and a Fiat out of wood. His works have been displayed in Paris; London; Tokyo, Japan; in *Vogue* and *Interior Design;* on BBC and MTV. He's used persistence and talent to carve himself out a unique position in the world of wood. And he does look like a guy who's got his shit together, a guy who's paid his dues, a guy who has a right to sell a divinely crafted pine trench coat on a pine coat hanger for 12,000 Euros.

As I leave, lacking the appropriate vocabulary, I give him the *Wayne's World* "I am not worthy" bow along with an "arrivaderci." He looks at me as if I'm from another planet. And compared with the world Livio works in, he's right.

WOODWORKING BLIND—JUST LIKE EVERYONE ELSE

Walk into Gordon Mitchell's, David Albrektson's, or Ron Faulkner's woodworking shop during the day, and you'll see the normal woodworking machinery—table saws, routers, drill presses, and planers. Walk into their shops with the lights off on a pitch-black night, and of course you won't see a thing. But you might hear something. You might hear the whir of a table saw, the roar of a planer, or the whack of a hammer. These three woodworkers are blind—and lights to them are about as useful as a rubber screwdriver.

What kinds of things do Mitchell, Albrektson, and Faulkner turn out? The same things as most other woodworkers. Albrektson currently favors decorative boxes with dovetails, finger joints, and mitered joinery. Though he's tackled projects as ambitious as three-legged corner tables with tapered legs and mortise-and-tenon joinery, he gravitates toward the uncomplicated, not because of difficulty but rather because of his preference for simple Scandinavian and Shaker-style design. Mitchell uses his woodworking finesse to tackle projects ranging from desks to whirligigs to outdoor furniture. Faulkner tackles "that which needs to be done," and that which has needed to be done includes fold-up dining room tables, bookcases, and kitchen cabinets.

They're not alone. John Cook, a blind cabinetmaker, has tackled scores of kitchen cabinets, bookcases, and desks. Stephen Naylor, who gained national notoriety when he was featured in television commercials for Sears power tools (so good you could use them blindfolded), builds custom furniture ranging from walnut rocking chairs to dining room tables to 7-foot-tall Federal-style armoires.[3] Not all the projects fit neatly inside of a shop. In fact, Mitchell, with the assistance of his thirteen-year-old daughter, built his shop—and a ranch house to go with it. He shied away from nothing, shingling, wiring, hauling up roof trusses. Faulkner made the doors, made the door and window trim, and laid the hickory hardwood flooring in the house where he now lives.

Each of these woodworkers make it clear that while they certainly face special challenges arising from their sightlessness, in the larger scheme of things they face the exact same challenges as do sighted woodworkers—challenges like safety, accuracy, interpreting plans and instructions, patience, and customer satisfaction. Albrektson explains, "Lots of times I'll make a project, and then a few years later after I've gotten better at a certain technique, I'll say 'Get that thing outa here.'" Just like everyone else.

John Cook, who at one time ran a cabinet shop employing thirteen people, belongs to the nine-finger club. "When I lost a finger, it was for the same reason any other woodworker lost one of his or hers—I did something dumb, something I knew very well not to do. In my case it had nothing to do with not seeing, but instead doing something stupid, such as back cutting with a molding head cutter on a table saw. Before starting the saw, I stood right there and said to myself 'you know not to do this,' but did it anyway. This just illustrates the fact that those of us with some physical disability are no different from anyone else. Severing a finger can be quite painful, but not nearly as painful as the devastation of one's pride."[4] Just like everyone else.

Faulkner outlines his two main challenges. The first is "learning how to think through your fingers" when going from sightedness to blindness. The second relates to the design process. Creating a mental blueprint of the project can be a hurdle, even though he feels he holds an advantage by having honed a keen sense of three-dimensional spatial orientation from his years as a commercial pilot. He talks about the difficulty of not only getting a mental picture of what the finished project will look like to him but, in the case of working for a client, the added challenge of divining the picture of what the *client* thinks it will look like.

And there are the roadblocks of overcoming misconceptions and stereotypes. "I get seen at the local Woodcraft supply, and I'm there with my cane, and I know there are people there thinking 'what's this blind guy doing? Maybe he's just out there making sawdust,'" muses Albrektson.

Measuring accurately is perhaps one of the greatest challenges.

Gordon Mitchell using a "1-2-3" block along with fractional measuring blocks to set the fence on his table saw. Not content with simply working in the shop, Mitchell built his shop and a ranch house to go along with it.

Sight-impaired woodworkers approach the problem in a variety of ways. The only tools made specifically for the blind that Mitchell uses are a 36-inch folding rule and a 12-inch metal rule with Braille markings. But beyond these, he uses a rag-tag arsenal of tools. He uses a carpenter's square with 24- and 16-inch-long legs, 2 inches and 1 1/2 inches wide, respectively, for some measurements. (The square came in particularly handy when he was building a house with studs, joists, and rafters set on 16- and 24-inch centers.) He uses 4-, 3-, and 2-foot levels for intermediate-length measurements and a standard tape measure with a locking mechanism for transferring longer lengths between projects and boards. For fine tuning he uses a set of aluminum measuring blocks that are 1/16 inch, 1/8 inch, 1/4 inch, 1/2 inch, and 3/4 inch thick, along with another block that is exactly 3 inches, 2 inches, or 1 inch, depending on how it's turned. They're handy for measuring, setting saw blade and router bit depths, and a variety of other tasks.

Cook, like Mitchell, uses measuring blocks, but his most useful measuring tool is a 4-foot-long stick that has precise kerfs cut into it every inch. It also has double kerfs marking each 1-foot increment. This allows Cook to slip the desired kerf over his table saw blade and add the necessary fractional blocks for setting his table saw fence. Cook explains, "After years of practice, my ability to feel minute inconsistencies enables me to see things with my fingers that you may not see with your eyes."

Albrektson uses all of the above and more: "Sometimes you just need to improvise when you measure. It might be adding together the thickness of a credit card, a matchbook cover, and piece of wax paper to get what you need. It's what works for the moment." And the wood? Albrektson can recognize maple, walnut, pine, and other woods he uses regularly by smell when he cuts into them and by touch by running a fingernail across them to feel the grain.

Faulkner favors a tool called a Rotomatic; a 6-inch-long rod with sixteen threads per inch and a square nut. Each full turn of the nut represents 1/16 inch, while a quarter turn measures 1/64 inch. Extensions measuring 6, 12, and 18 inches, screwed onto the end, can give the measuring device a reach of 42 inches. A similar device—a click ruler—is a rod that audibly clicks every 1/16 inch as it's adjusted, has tactile markings, and can measure distances up to 4 feet. Faulkner also uses his hand, which measures 9 inches from tip of thumb to tip of middle finger, to step off rough distances.

Cutting and shaping the wood presents another set of challenges. But since most of the stationary tools used—shapers, table saws, miter saws—are machines with fixed guides, once the fence is set correctly, cutting or shaping the wood is actually pretty straightforward.

Mitchell expands on the beauty of using fixed guide machines. Since most are permanently positioned in one spot, he knows exactly where they are in his shop, exactly how to set them up, and exactly where the accessories are located. As contrary as it might seem, he avoids using push sticks when cutting lumber on his table saw. Using his hands allows him to feel the wood for any movement or rise. And he listens for changes in the speed and pitch of the motor and blade that might foreshadow binding or kickback. Faulkner likes heavy-duty machines, like his massive Powermatic table saw, jointer, and shaper, which he knows aren't going to move, tilt, or tip and leave him in an uncompromising position.

Mitchell explains that he *can* learn how to build a project by running his hands over a completed version of it, but he also likes to learn from printed matter. However, most woodworking magazines and books are very visual, presenting an obstacle for those like Mitchell who seek new projects and techniques. Mitchell, a retired computer programmer and analyst, has software programs like JAWS (job access with speech) that will translate text files into audio, but in the eyes of the software, the graphics don't exist. And there's no such thing as skimming the text; you wind up listening to every single word. His wife can explain the content of some photos and illustrations, and e-mails to authors of articles can yield additional information, but often that's not enough.

Which is where Larry Martin and Woodworking for the Blind come striding in. Martin, hearing of Mitchell's dilemma, volunteered to be his eyes. And from the whirligig project from which it all started, the organization grew. Via a combination of podcasting, burning CDs, recording with the Macintosh GarageBand program, and iTunes, Martin has released ten CDs containing recordings of twenty-eight magazine issues, of which over two dozen blind or sight-impaired woodworkers now take advantage. The recordings—a hun-

Ron Faulkner with his homemade kiln, which will dry 100 board feet of lumber per load. Larry Martin of Woodworking for the Blind helped Faulkner by verbally explaining the details from the illustrations and photos published in a woodworking article.

dred hours' worth at this point—include the full text of articles along with general descriptions of the photos, illustrations, and plans.[5]

"The stuff that Larry's doing has been one of the most meaningful things to me—ever," says Albrektson. Faulkner is equally appreciative. "I wouldn't have my wood kiln today if Larry hadn't taken a weekend to explain the details from a couple of articles." A community where blind woodworkers can share their tips and questions via e-mail is now evolving.

As with anybody in any endeavor, the triumphs of these unique woodworkers outweigh the negatives. Albrektson explains, "Initially when I made mistakes I would become enraged, but now I don't expect to get it right the first time out. Now I enjoy the process. To me, making a good mortise-and-tenon joint can be very satisfying." Mitchell concurs: "I won't stack the quality of my work up against the quality of yours. I do what I do because I enjoy it, not to prove a point."[6] He adds, "If my work pleases my wife, what more can I wish for?" Not all of Mitchell's projects turn out flawless. "My wife's tole painting has come in handy, covering up my mistakes."[7]

One posting by a blind woodworker on the WOODWEB Web site adds to the overall sentiment. "I actually have found that I do better work now that I am blind. When I was sighted I would rush projects sometimes, and things would not always turn out the way I wanted. Now that I am blind, I have to plan everything down to the smallest measurement before I do anything. With the mental plans and my wife as my shopper at the hardware store, I waste less than when I was sighted and end up with much better finished products. Maybe blindness has just made me a much more patient craftsman."

HOW MUCH WOOD WOULD A WOOD COLLECTOR COLLECT?

You say you have an unusual collection, like David Morgan with his 137 traffic cones? Or you're a hardcore birder with 1,738 sightings on your Life List? Or you've endured some serious sideways glances while showing off your collection of 3,728 airline sickness bags?[8] Well, folks, it's time to start taking your hobby a little more seriously. Because compared with Gary Green you're a lightweight, a ninny, a mere dabbler in your avocation. For Green has encountered alligators, snakes, and needle-shooting cacti in pursuit of *his* hobby. He's spent ten of his most recent ten two-week vacations traveling to forty-eight states, logging thousands of miles. He's slid off mountains, been caught trespassing, and walked 6 miles after slashing a tire in the middle of nowhere.

And what does he have to show for all this? Thousands upon thousands of pieces of wood, each about the size of your hand, each smoothly sanded, each different. Welcome to the world of wood collecting.

For Green, his xylarium, or wood collection, all started innocently while he was splitting firewood one day in 1992. He began wondering what the woods would look like planed down. By the end of the day he had ten nice small samples. "People started bringing me pieces of firewood from trees in their yards. Then I got a bit more serious and started collecting Indiana natives [trees] in the wild." His collection grew to fifty, then a hundred, eventually twenty-five hundred pieces. With the purchase of the collection of a fellow enthusiast who'd gone to the big xylarium in the sky, his total jumped to six thousand-something. He won't know the exact number until he's finished cataloging the collection and eliminating duplicates. "I'll let you know in about ten years," Green explains.

Green is not alone in his passion. The International Wood Collector's Society (IWCS), founded in 1947, has over eleven hundred members in fifty states and thirty-five countries, including Estonia and Iceland. Its bimonthly *World of Wood* magazine is a glossy four-color affair with article titles like "Botanical Adventures in Mayan Zinacantan," "Using Odour in Wood Identification," "Strzelecki Gum in Neerim Finds Some Friends," and "Scientist Report: Figs Were First Crop." In the IWCS directory you'll find levels of enthusiasm indicated by e-mail addresses like mrwood96@, sawdust101@, woodturner@, woodgal@, galleryofwood@, desertwood@, woodchipper1@, ironwood@, and not one but two tallwood@. It's a friendly lot; you'll find a small house next to the names of those willing to host traveling IWCS members on the prowl for wood.

It's a bottomless hobby. With as many as eighty thousand species of trees on the planet, a person gathering one wood a day would still require 219 years to amass a complete collection. Even Richard Crow of England, who has what most consider to be the world's largest private collection, has *only* (only!) seven thousand species; less than 10 percent of the woods nature has given us. This from a man who has collected for fifty years; who has traveled to the far corners of the earth, including the war-torn republic of Georgia; and whose family has been in the lumber business since 1795. Given the enormity of the task, most collectors use a varied approach; they'll field-collect some (create their own from scratch, like Green from his firewood), buy some, trade some, and cut some out of larger boards.

Raimund Aichbauer's collection contains thirty-five hundred identified wood samples, four hundred curiosities, and seven hundred microscope slides. A member of the Dutch Society of Wood Sample Collectors, he's been collecting for twenty-five years.

Collectors have a wide range of MOs and specialties. Steve Bartocci, who recycles wood and sells it to woodworkers, finds exotic woods like Borneo rosewood and blonde ebony in pallets and crates used to ship goods into the United States. And while using rosewood for pallets may seem akin to using a Stradivarius to play "My Sharona," Bartocci explains, "Exotic wood is generally not exotic in its country of origin." Another collector has gotten his share of rare woods from the dumpster behind the Martin guitar factory. Alan Curtis, a retired forester and the man considered by most in the IWCS to be *the* ultimate wood connoisseur, has stomped the forests, jungles, and deserts of nearly twenty countries and has field-collected over seven hundred species. The Internet has made once nearly-impossible-to-obtain woods accessible. Green can ship you eighty-two different woods, most with the modest price tag of $1 to $3—and those are just the ones starting with the letter A. Tim Heggaton offers more than a thousand different species.

Though standard specimens 1/2 inch by 3 inches by 6 inches are little bigger than a Hershey bar, large collections take up a cumulatively large amount of space. Green began storing his collection on the walls of his office, ran out of walls (and discovered that samples faded in sunlight), and now stores most in plastic tubs. Bill Mudry, who lives in a one-bedroom apartment, stores his 2,130 samples in five chests of drawers. Peter Damm, a guitar maker from the Netherlands, has his 1,500 samples alphabetized and organized—neat as a pin—card catalog style. An English collector who lives in a small flat collects specimens as cross-sections on microscope slides. Some display their samples like books on a shelf. Manuel Soler built a shed in his backyard to shelter his collection. Ralph Cox echoes the words of many in saying, "My wife thinks I have too much wood, but I really don't."

Many IWCS members are woodworkers and choose to display their collections as objects they've created. What kinds of objects? Carved wombats, forests of 2-inch-high trees, bowls, vases, rice-size goblets, plates, swans, quilts, miniature spiral staircases, pens, spoons, miniature rolling pins, lapel pins, whittled boots, carved thumbs, boxes, and thimbles. One collector has whittled two thousand dolphins; another has turned eleven hundred different eggs from eleven hundred different species on his lathe. Maps with countries, states, or even counties crafted from woods native to that area are popular. One member, advertising in a recent issue of the *World of Wood*, expressed his need for a small piece of Socotra cucumbertree or dragon's blood tree to complete a world map.

Some collectors can lay claim to the title of artisan. Norm Sartorius has been making fanciful whimsical spoons for over thirty years and has works in the Smithsonian. He's carved his creations from bogwood, anonymous desert roots, and deceased people's furniture brought to him by sentimental relatives. He explains, "Some people are emotionally connected to wood, a tree, a piece of their home, a family legend, an experience in the presence of a special tree. When something is made for them from that wood, then the emotional content is retained and the object then becomes a trigger for a memory."[9] Allan Boardman makes wooden puzzles—not just any wooden puzzles but miniature three-dimensional ones like his interlocking nineteen-piece Great Pagoda, measuring half an inch across.

Labeling samples is a must. Official names are written in the uni-

Wood collector Allan Boardman likes to create miniature puzzles from some of his specimens. This "Great Pagoda" puzzle contains nineteen pieces, yet is a mere 1/2 inch across.

versally accepted language of science: Latin. Labels always consist of two parts: the genus name (first letter always capitalized), followed by the species name (first letter always lower case). If used in a scientific publication, that name will be followed by the initial or name of the person who first named the plant. When you see a board labeled *Quercus alba L.*, let there be no doubt you're looking at white oak, first described by Linnaeus. Many collectors include information on the who, how, what, when, where, and why of the wood in a database. Bill Mudry's research file stands at an impressive 3.5 megs.

There are war stories. Bartocci explains, "As a kid growing up in New Mexico I used to collect spiders, tarantulas, and scorpions and never got bit or stung. I found it ironic that as an adult living in a rainy Seattle, while cutting a large hollow log of Arizona ironwood, I was bitten by a black widow spider who called the log home." Brian Baker, who turns saucers out of specimens—eight hundred, all different species—explains "Most [woods] were pleasurable, with good aroma, turnability, and polish, but there are some I hope to never turn again, like *Toxicodendron radicans*—poison ivy." Satorius, who teaches an occasional woodworking class, has a different brand of chilling story. He once told his students to bring in a piece of wood that meant something to them. An older gentleman brought in a piece of wood from a hanging tree. "Some people in class would not even touch the wood," Satorius explains. "It was a plain piece of oak, unremarkable in color and grain, but laden with history of a rather grave nature."

While this all may seem to edge upon fanaticism, it's no different from any other type of collecting. Whether it's Beanie Babies, baseball cards, or traffic cones, collecting isn't so much about collecting things as about collecting experiences and memories. Green explains, "When I discuss my trips with friends and acquaintances, many question my mental well-being. But there is a real satisfaction to pulling a specimen from my collection and recalling the adventure that came with locating it."

Collectors can (and some do) amass gigantic collections without ever setting foot in a forest or emptying a mote of sawdust from their cuffs. One can order a sample of exotic wood, 1/2 inch by 3 inches by 6 inches, from the deepest jungles of Central America over the Internet for the price of a latte. But that's a pretty bleached-out memory compared with that of Manuel Soler of Spain, who can pick up *his* sample of the same species and retell the story of running in terror from millions of man-eating ants in a rain forest in Costa Rica while obtaining his specimen. *That's* a memory. *That's* a collection.

If you think these collectors have a daunting array of woods, walk into the Forest Product Laboratory (FPL) in Madison, Wisconsin, and ask to take a look at *their* wood collection. There you'll find not dozens, not hundreds, not thousands, but tens of thousands of wood specimens—a hundred thousand, to be exact.[10]

Alex Wiedenhoeft, an FPL botanist who needs only to lean back in his chair and peek out his office door to see all hundred thousand specimens, ruminates: "I worked here for a long time thinking this was the dullest place on earth, because basically the wood collection is drawers and drawers of little blocks of wood. It's the world's largest wood library, and like a regular library, if you don't know how to read, it *is* the dullest place on earth. It's not until you start reading the blocks that you realize what an amazing place it is. The more you learn about this stuff, the more interesting it is."

The FPL wood collection, born in 1910, was—oddly enough for its day—initially headed by a woman, Eloise Gerry. "I must admit the Forest Service did not want a woman, but as it happened there wasn't any man willing to come and do the work," she explained.[11] Gerry grew the collection by obtaining samples through expositions, colleagues, collection acquisitions, and foresters. Oversight of the FPL collection was dominated for the next seventy years by a succession of men with long tenure. Arthur Koehler joined FPL in 1914 and stayed for thirty-four years. Bohumil Kukachka (Kuky to nearly everyone) became curator in 1945 and remained through the early 1970s. Regis Miller was hired in 1970 and retired only recently.

Kukachka's hiring is legendary, though legendary only to inveterate wood geeks. During his interview for the position of wood anatomist, Kuky casually plucked three unmarked specimens of wood off Koehler's desk and proclaimed, "Ah, Chilean woods. *Nothofagus, Laurelia,* and *Aextoxicon.*" He was hired on the spot.

The wood collection grew in fits and starts. At the start of Koehler's tenure, the collection stood at only a few thousand samples, mostly domestic. By the time he left, the collection stood at about eleven thousand, though there are indications it was a *messy* eleven thousand. Though perhaps not the best housekeeper, Koehler did put the field of wood anatomy on the map by helping solve the crime of the century, the Lindbergh kidnapping crime.

When Kuky took over, he cleaned house, throwing out as many as six thousand samples that were without proper documentation, and organizing the collection alphabetically. He standardized the specimen size to 80 × 100 millimeters to conveniently fit into a platoon of custom-made wood drawer cabinets, and he instituted an index card filing system. By the time Miller stepped on the scene, the collection had more than tripled in size. This was due in part to the acquisition of the fifty-five-hundred-specimen Samuel J. Record (SJR) wood collection from the School of Forestry at Yale, the world's largest collection, which, in the 1960s found itself with no staff, no budget, and no one particularly interested in maintaining the collection. The collection included the critical herbarium sheets: samples of the leaves, seeds, flowers, and other "soft" material of the trees from which the wood was harvested. (Herbaceous material is critical, since the woods of many trees are anatomically identical, and true differentiation can be determined only by leaves, seeds, and so forth.)

The Yale collection was arranged numerically, which, according to Miller, "was convenient, but if you needed to look at a certain wood you might have to pull out sixteen drawers in sixteen different places to look at sixteen different specimens." One of his first tasks at the FPL was to rearrange the woods by family and genus, a task he describes as "dusty, dirty, tedious, and time consuming." It took him three years. The Yale filing cabinets—and much of the wood within—were of a different size than the Madison collection, so in the end it was decided to keep the collections separate. They remain separate today, and, since new finds are integrated into the FPL collection, the SJR collection remains basically petrified in time.

In the early 1970s, the Field Museum of Chicago bequeathed its collection of twenty thousand specimens—twenty thousand chaotic specimens—to the FPL. It came in moving boxes, with no cabinets, files, index cards, or herbarium. Kuky integrated the Field Collection into that of the FPL, but not before examining every single sample with a hand lens to confirm its proper identification. Miller explains, "If the sample had been misidentified, it was discarded. In addition, [Kuky] discarded many samples that were not vouchered, although I sometimes retrieved specimens from the trash if they seemed to be of historic significance or in short supply." By the time the dust settled, fewer than half of the specimens were incorporated.

Specimens from dozens of other specialty collections and collectors have been integrated into the FPL collection over the years. The largest is the Krukoff collection, featuring 3,168 woods, mostly from South America. Others include the DeWitte collection, with 249 woods from the Congo and Zaire; the Fors collection, with 311 woods from Cuba; the Detienne collection, with 202 woods from the Ivory Coast; and the Stahel collection, with 355 woods from Suriname.

Scattered about the FPL offices are dozens of mammoth tree "rounds" from the Jesup collection. They're much (much) larger than the standard-size wood sample but not nearly as large as they were originally. "In the 1860s, this guy had the foresight to cut 5-foot-long boles from the trunk of every type of tree that grew in the United States. And he got 'em all." explains Miller, wearing the omnipresent 10× magnifying loupe around his neck. "Back then you could just chop down whatever you wanted. Today, even if you could do that, it would cost you a fortune." The American Museum of Natural History housed the collection until 1964 but then, running out of space, offered the exhibit to any museum that could house it. The World Forestry Organization in Portland, Oregon, became caretaker for a while but also ran out of room. The entire collection of five hundred weighed 200,000 pounds and filled two box cars. The FPL cut and saved end rounds from many of the boles, but no other institution or person had room to store them all. It's rumored that the irreplaceable boles are now homeless, perhaps in Tennessee.

Over the past few decades, the specimens of the FPL collection have been entered into a database system, accessible via the Internet. There you can find 45,000 specimens representing 14,000 species. One example of the comprehensiveness of the collection is *Quercus*—known more commonly as oak—represented by 222 species. Click a few buttons and you'll encounter common names: white oak, brown oak, red oak, and live oak; descriptive names: prickly-leaved silky oak, bootlace oak, blush tulip oak, thready-bark oak, and dwarf post oak; and bizarre names: mel-of-Jerusalem oak, Harvard shin oak, wainscot oak, poak, and moak.

So how much wood *could* a wood collector collect? Keep counting.

NAKASHIMA: THE PAVAROTTI OF WOODWORKING STILL SINGS

Ask woodworkers around the world to name the most influential woodworkers of the twentieth century, and you'll hear the names Tage, Sam, Wendell, and, well, Norm a lot. But another name you'll hear over and over again is George. George Nakashima. Perhaps it's because George Nakashima—born in America of parents of Japanese ancestry, educated partly in France, a onetime member of an ashram in India—considered himself a world citizen. Perhaps it's because Nakashima, as an architect, designer, father of the American Crafts movement, and peace ambassador, cast a wide net. Perhaps his fame is even due in part to the hubbub over the increasing collectiblity and value of his work; at a recent auction an English burl oak table he made in 1973 sold for over $200,000.[12]

But Nakashima isn't known only for the things he *did* do, but also for the things he *didn't* do. He didn't contort trees into things they were not. He didn't slice mighty logs into anorexic boards and wimpy veneers. He didn't consider a burl, crotch, or knothole something to be discarded, but rather something to be celebrated. And though George Nakashima died in 1990, the George Nakashima Woodworker studio didn't. Mira Nakashima-Yarnall didn't allow it.

I arrive at the Nakashima studio in historic Bucks County, Pennsylvania, flustered after a frenzied drive that took me from LaGuardia Airport through the heart of Manhattan and on to the manic New Jersey Turnpike, where, upon realizing too late I was in the EZ-Pay lane with trucks blasting their horns behind me, I ran the toll booth gate, setting off alarms, cameras, and a still-lingering paranoia that at any moment a state trooper could come lumbering toward me, handcuffs in hand.

But I find an immediate calmness in the presence of Mira. She is gracious, wise, and quick to laugh. She's been at the helm of the studio for sixteen years. But she didn't just hop into the chair vacated by her father; she designed and crafted one of her own, then slid it next to his. She's earned her place at the head of the table.

When asked if she senses her father's spirit is still working here, Mira smiles. "I think that's the reason I'm still doing this. Somehow by keeping the wood going it's as if Dad were still alive here, and that's what keeps *me* going."

To get the full story, one must rewind a hundred-plus years. George Nakashima was born in Spokane, Washington, in 1905, to parents descended from samurai. He attended the University of Washington, the Ecole Americaine des Beaux-Arts in France, and graduated with a degree in architecture from MIT in 1930. Shortly afterward he began an around-the-world tour by steamship, living in Paris, Tokyo, and an ashram in India.

He studied the architecture of Le Corbusier (from whom he gained a lifetime fascination with, ironically enough, reinforced concrete) and Frank Lloyd Wright (to whom he was grateful for showing him what *not* to do with his life).[13] For two years he absorbed the teachings of Sri Aurobindo, which included the concept of karma yoga, the path of action.

He spent a year interned in a relocation camp in Idaho during World War II. Here he had the opportunity to work alongside, and learn from, a fellow internee well versed in the art of traditional Japanese carpentry. He was allowed to leave in 1943 and, with his wife, Marion, and infant daughter, Mira, moved to Bucks County, where they eventually settled on a 9-acre plot of wooded land five miles from the spot where Washington crossed the Delaware. This small plot of land, which eventually sprouted fifteen buildings of every size and shape, is referred to by the family and workers as the complex.

Nakashima labeled himself a Hindu-Catholic-Shaker-Japanese-American—with Druid sometimes thrown in for good measure. This early tangle of people, travels, and experiences helped forge George Nakashima's philosophy toward wood, design, and life. And in the end you discover that a piece of Nakashima's furniture, like its creator, has a simplicity on the outside, driven by a complexity within.

My morning with Mira progresses through the complex in the same manner as a piece of furniture. We begin in the wood shed, progress through the workshop areas, move on to the finishing area, and wind up in the Conoid Studio, a sort of office-museum-history center filled with a casual retrospect of finished furniture.

Calling Nakashima's inventory of wood a lumberyard is like calling the Hope Diamond a pebble. All the storage sheds combined contain in the neighborhood of 180,000 board feet of lumber. Much of the lumber is stacked in boles, logs that have been sawn lengthwise—through-and-through—into boards, then restacked, with spacers between, in their natural order. This maintains the wholeness of the tree and allows a single tree to be used for a matched set of chairs or a massive table. Traditional quartersawing—where cuts are made on the radius with grain running perpendicular to the surface—may produce boards that are straight-grained and stable. Mechanized rotary cutting—where thin layers are shaved from a spinning log—may produce miles of veneer. But cutting through-and through—also known as plainsawing—in the Nakashima tradition produces planks that reveal their entire range of graining, figuring, and color. It exposes the intersections of trunk and branches, crotch structure, and one-of-a-kind acts of nature.

Mira Nakashima standing before a humongous wood slab in the storage shed at the Bucks County, Pennsylvania, compound. Among this vast supply of wood—some 180,000 board feet—are still slabs of wood purchased decades ago by her father, George.

You'll find few thin boards and zero veneers in the Nakashima wood archives. What you will find are 2-, 3-, and 4-inch chunky planks of wood with bark still intact. There are massive slabs of claro walnut 15 feet long and 6 feet wide, figured bubinga from Africa, English oak burl, eastern black walnut, Persian walnut (actually harvested when Persia *was* Persia), figure and crotch walnut, cherry, redwood roots,

Oregon myrtle, maple, and East Indian laurel, some of it dating back to when Nakashima himself acquired it. One shed is used to store the most massive of the massive slabs in an upright position. I click a picture of Mira standing before one stack of slabs to give a sense of scale; the boards are twice her height and twice her width.

Most of these great slabs are destined to become solid wood tables, benches, and other pieces of furniture. In *The Soul of a Tree*, Nakashima explains, "The tree's fate rests with the woodworker. In hundreds of years its lively juices have nurtured its unique substance. A graining, a subtle coloring, an aura, a presence will exist this once, never to reappear. It is to catch this moment, to identify with this presence, to find this fleeting relationship, to capture its spirit, which challenges the woodworker."

The trees and slabs have been cut purposefully and sensitively. "Trees have a yearning to live again, perhaps to provide the beauty, strength, and utility to serve man, even to become an object of great artistic worth. Each tree, every part of each tree has only one perfect use."[14]

Perhaps the most perfect of perfect uses Nakashima found for his mighty slabs was creating his Altars for Peace. In 1984, he acquired the largest book-matched walnut boards he'd encountered in all his days of woodworking: each 3 inches thick, 6 feet wide, and 12 feet long. Awakening from anesthesia after surgery, he had a vision to create the most spectacular piece of furniture ever made, something that would represent the positive spirit of both nature and man. If this wasn't ambitious enough, Nakashima decided to make six altars, one for each continent (sorry, Antarctica). His first was installed in the Cathedral of St. John the Divine in New York City in 1986. Two more Altars for Peace—their creation guided by Mira—have been installed since his death. In 1995, one sawn from the same log as the New York altar was installed in Russia. In 1996, a third, cut from a tree of nearly identical dimensions, was installed in India. Each is as immensely beautiful as it is immense. Wood for the final three is on hand.

* * *

The studio has an informal network of scouts and sawyers who make available woods from around the world. Nearby, Talerico and Hearne Hardwoods supply woods from England, Germany, Central America, and other parts of the world. Another sawyer on the West Coast keeps them stocked with redwood burls. But much of the wood is obtained locally, sawn by John Kirlew, who was the assistant of Frank Kozlowski, George Nakashima's logger and sawyer for decades. Mira speaks with mixed emotions about the abundance of local walnut logs they've been able to obtain. "We've been very well supplied with big logs lately because of all the housing developments," she says. "They take down all the trees, especially the walnut trees. Nobody seems to want them on their property, so we've been lucky."

They haven't been lucky in all respects. "It's been a bad year for lumber," Mira explains. "We discovered powderpost beetles up in the loft where Dad kept some of his lumber. The wood had been up there so long that the bugs kind of took over. We had to fumigate all three buildings." The bright side of the near disaster was that most of the wood escaped serious damage and it gave Mira and crew a chance to inventory, rediscover, and reorganize their inventory. Many boards had been lying in wait for decades, biding their time for just the right project to come along. Workers are in the process of tagging all the slabs and planks, and putting the information on computer, so when they're looking for a species of a specific thickness and size they'll know where to find it. Mira expects it will take years to finish the task.

We take a short stroll to the woodworking shop. Contrary to what one might expect of a studio legendary for its handwork, the building contains a full array of industrial-duty table saws, jointers, bandsaws, and planers. For Nakashima, this was not a disconnect. He felt that as long as man controlled the final product, using modern machinery was fair game. But you won't find computer numeric control routers, lasers, or other weapons of mass production. With the exception of some chairs, nearly everything is one-of-a-kind or custom-made. Once machines have roughed out the pieces, it's handwork that carries the piece to the finish line.

We walk past a 2-inch-thick, 4 × 6 foot natural-edge maple burl slab that's been flattened and ready for dressing down. "You really can't see what it looks like when it's rough," explains Mira. "You see a little more when you clean it up, but you really don't see the color and texture until you get through the entire process." But even to my unpracticed eye, I can tell this is going to be drop-dead gorgeous.

It is natural-edge wonders like this for which Nakashima is known. Mira tells the story of the style's genesis, a story that she herself has just learned from one of her father's early suppliers. Nakashima was visiting Thompson Mahogany in Philadelphia, a company that purchased only pure, flawless logs for cutting the finest veneers. Those with knots or wild grain were rejected. Nakashima struck a deal with the owner to buy the rejects. For the first batches he traveled to Virginia to oversee the milling so the logs would be cut not into anonymous square-edge boards but into unique slabs complete with crotches, burls, and warts. When the first load of raw slabs arrived at Thompson, accompanying the flawless veneer logs, the foreman nearly threw the stuff out. "No client would want that junk," the foreman proclaimed. But that is how Nakashima got his start working with natural edge slabs—with rejects.

He respected each tree for its unique color, shape, and personality. Instead of squaring off the edges and ends like everyone else, Nakashima left the wood in its natural state. He listened to the wood, then proceeded to use the unique voice of each tree to guide him in creating the fabulous furniture for which he became known. Initially, people thought Nakashima was crazy for creating furniture punctuated with holes, cracks, and butterflies. Eventually, they began paying him top dollar for it.

The aforementioned butterflies are the bowtie-shaped inlays that evolved into another Nakashima trademark. Butterflies had been used for years by Shaker, European, and Japanese woodworkers as a form of concealed joinery. But Nakashima began showcasing them—using them as decorative accents as well as mechanical devices to hold boards or halves of a cracked slab together. Then as now, the selection and placement of a butterfly required taking into account size,

A natural-edge table with trademark Nakashima butterfly inlays in the Conoid House. The woodworking studio continues to offer furniture designed by both generations of Nakashimas—but if you order a piece, be prepared for an eighteen-month wait.

color, direction, contrast, expansion, and contraction. The first butterflies were inlayed using only hand tools, but early on Nakashima began using a router for the initial cutouts. Nakashima writes, "In a creative craft, it becomes a question of responsibility, whether it is man or machine that controls the work's progress."

A mammoth Timesaver—a 4-foot-wide belt sander used to

smooth large slabs—standing in one corner of the shop is symbolic of the changes that have occurred over the past decade. The machine takes a task previously done by hand and produces faster and more consistent results, while freeing up workers to pay greater attention to the details that truly require the craftsman's eye and hand. Nakashima Woodworker isn't locked into the past; if a process needs improvement, it changes.

Jason, in the chair shop a short walk away, is a living example of this. He's recently created a new system for cutting slots in the legs, which has solved a recurring problem of the feet not running parallel to one another. It took trial and error, but a better joint was created. If George were standing next to him, he would approve. "Good joinery, whether in buildings or for furniture, is difficult to design and even more difficult to execute. It should be thought of as an investment, an unseen morality."[15]

Jason discusses the informal system used to pass down woodworking methods from one generation of woodworkers to the next. "You get the basic rundown from the people who did the job before you, but if you find a better or faster way to do it, you do it. There are as many ways to do things as the stars in the sky. Everyone approaches things differently. Everyone one has their own 'Well, that looks right to me.' The guy who worked here before me liked things a littler rounder. Mira likes things a little more angular; her father did as well, and so do I."

"That's what gives each piece personality," Mira adds.

The chair spindles are turned on a lathe, but after that, each is planed by hand. The grain is "read," and each spindle is faceted, giving it a unique appearance depending on how light hits it. "It's a different feel than a sanded spindle," explains Mira. "It's almost a throwback to traditional Japanese carpentry. They don't sand anything; they use planes."

We encounter a tangible example of how the old influences the new as we head out the door. Mira spies a stool in the corner that bears her namesake: the Mira Chair. When asked why it's there, Jason responds that he's using it as a model for a recently placed order; he wants to make sure he gets the details right. The stool is over fifty years old. It is *the* stool featured in a *Look* magazine article from 1952: “A Chair for Mira.” One photo shows George Nakashima standing behind a burl oak slab. But it’s a photo of a beaming ten-year-old cowboy-boot-clad Mira peering from behind the chair that dominates the page.

One cannot help but notice that the atmosphere is unusually serene for a place that employs industrial planers, 500-pound slabs of wood, and a workforce of fifteen. Workers aren’t listening to iPods, they’re listening to the wood. There is a sense that people are given the time to do the job right, and given that, they do the job right.

Another small building houses the finishing department. This is one place where Mira would prefer that things remain the same as they were fifty years ago. But the products have changed—for the worse. They’ve been getting some complaints about water spotting on dining room tables. The studio has always used three to five coats of hand-rubbed natural-oil finish to bring out the depth, color, and grain patterns. But the composition of the oil is now somehow different. They’ve pondered switching to polyurethane, which would repel water better, but it would bury the grain, place a hard barrier between wood and caressor, and make scratches more difficult to repair. So they continue to tweak their linseed oil formula. And they listen to the echo of George Nakashima: “the finest finish of all can sometimes result from simply aging, like the finish of the timbers in Japanese wood temples, some of which are over a thousand years old.”[16]

We spend the next hour in the Conoid Studio—an arch-shaped thin-shelled reinforced concrete structure built by Nakashima in the 1950s. It has an open yet intimate feel. The building inspired Nakashima’s Conoid line of furniture, and while some of the furniture shares the curvaceousness of the structure, the main design elements they share are solidity, simplicity, and grace.

The space is, in essence, a retrospective of Nakashima’s work. Visible are a classic natural-edge dining room table, massive freestanding slabs that serve as room dividers, and dozens of other pieces, some of which are one-of-a-kind prototypes. As I sit in one Conoid chair I realize they are not only comfortable to look at but supremely comfortable to sit in. Even the cherry and walnut flooring is spectacular. Mira tells me that some of it was crafted from wood that wasn't good enough for furniture, but she suspects some of it wound up as flooring because her father ran out of space for storing wood.

Mira is moderately irked by the number of Nakashima imitations—"knock-off-ashimas," she calls them—sold today. In her recent book, she explains, "Even though his forms are now imitated by many, the sincerity, honesty, and integrity of his work ensures that the work that follows it is at best a physical copy, unable to approach its essence. The unspoken thought processes and spiritual backdrop behind his creativity give his work an indescribable, elusive quality that is distinctively his."[17] That elusive quality lives on today at George Nakashima Woodworker. But it didn't live on without a struggle.

"We had a real slump after Dad died in 1990 because there was this myth, perpetuated by the press, that Dad built everything with his two hands. Even at the funeral downtown, the priest said, 'Well, there won't be any more furniture, now that these hands are still,' and I felt like waving my hands and saying, 'Hey, wait a minute; there's still a pile of us on the hill and a pile of orders to fill and a pile of wood.' And there was the myth that since the artist himself obviously was not around to sign his work anymore, it was worthless; that [the furniture] couldn't possibly be as good as when he was alive. So our orders disappeared."

Business became so lean that Mira began fearing if an order didn't come in on Saturday, there would be no work the next week. But the business eventually gained traction. An article titled "Against the Grain" in *Lear's* got things rolling. In 1993, the Michener Museum in Doylestown asked Mira to design the interior and furniture for a George Nakashima Memorial Reading Room. This produced what, in Mira's words, was "an almost embarrassing amount of publicity." Then, in 1998, the owners of the Moderne Gallery in Philadelphia invited her to participate in a joint exhibition and asked her to produce a new line of furniture of her own design. Which she did with aplomb—for while she may be Nakashima's daughter, she is also herself a woodworker, designer, and architect.

She graduated cum laude from Harvard University, earned her master's degree in architecture from Waseda University in Japan, has designed homes, and has had major furniture exhibitions in New York, Tokyo, and Boston. For the Moderne Gallery exhibit, she created the Keisho Collection—the Japanese word for "continuity"—which, according to the brochure "[continues] the evolution of design solutions, which preserves the methods and techniques embraced by her father." The designs are within the spirit of her father, yet different, lighter, perhaps more feminine. In comparing his mother's work with that of his grandfather, Mira's son, Satoru, told her it looked more like spaceships.

She worked alongside her father throughout her life, intensively for the twenty years preceding his death. There are photos of Mira at age three, sitting on her father's workbench drilling holes; at age ten, working alongside him in his shop; at age twenty-four, making small tables. This new branch of Nakashima Woodworker isn't something that's been grafted on; it's a natural part of the tree.

A good gauge of a business's health is the backlog of orders; right now, if you commission a Nakashima chair, be prepared to wait eighteen months before sitting down. They offer furniture in the classic George Nakashima designs—a Conoid coffee table for $6,500, a Minguren desk for $10,000, a music stand for $5,000. They offer Mira's newer Nakashima-inspired designs as well—a book-matched walnut dining table for $10,500, a lectern for $4,000, an arboretum bench of ipe for $8,800, and more. It may seem a high price to pay, but so artistic are the pieces in look and feel that one insurance company insures them as works of art rather than as furniture.

Father and daughter alike share a fondness for naming furniture after those that inspired the piece. There are Mira chairs, Mira mirrors, and Kevin end tables that George Nakashima named after his children. The Marion tea cart was named after his wife. Mira offers pieces with names such as the Maya table, named after her granddaughter, a Noguchi coffee table named after the designer, and a Michener coffee table named after the writer.

Mira flat-out admits that the future of George Nakashima Woodworker is uncertain. Mira and her husband, Jonathon Yarnall, who continues to work in the business, have four children, but none seems inclined to carry on the business. She talks about someday perhaps turning the entire operation over to the Nakashima Foundation for Peace. Another concept is to set up a woodworkers' collaborative, which would teach and perpetuate the Nakashimas' ideals and methods of work. But for now, the work and the passion continue.

In Mira's words, "No one does this work for money; the only motive that brings one to it is a sincerely felt passion for the wood and for the integrity of the entire design process. Perhaps we who work in wood are all a bit crazy; indeed, the loggers and sawyers I have known are as nearly crazy as I am. Indeed, we all become ecstatic at the sight of a beautiful old tree, especially when it is first opened up as lumber. It is this inspired insanity that keeps us going, and I often see the same passion overtake our clients when they prowl the woodshed in search of their ideal piece of wood."[18]

When I ask Mira what one thing she would like to convey to woodworkers, she defers to her father. "I think my father's contribution to the woodworkers of the world was that he didn't consider wood to be a material just to be made into something man wanted to make, but that wood had a voice of its own."

The Pavarotti of woodworking still sings.

MY ALMOST-PERFECT INTERVIEW WITH WOODWORKER JIMMY CARTER

In a request that has the same probability of being fulfilled as a flipped coin does of landing on its edge, I've written the Carter Center requesting an interview with the thirty-ninth president of the United States. The coin lands on its rim, and four weeks after sending off the request, I get an unexpected call from an assistant saying, "Mr.

President Jimmy Carter in the fully equipped workshop he's maintained in his two-car garage at home since leaving the presidency. Though he has a full array of power tools, he loves the historic, tactile feel of hand tools. One recent creation—a tall cabinet of persimmon—was auctioned for $1 million in a fund-raiser for the Carter Center at Emory University.

Carter has jotted 'twenty minutes OK' on the top of your note. We'll contact you in a few weeks to set something up."

To prepare, I read interviews, biographies, autobiographies, and things written about and by him. In the days before the scheduled phone interview I find myself on needles and nails. But once we're connected, and we talk through a few pleasantries, I feel comfortable—as if I'm talking with just another guy who likes wood. And I am. When I run through some of the topics I'll be covering in the book—the *Spruce Goose*, Stradivari violins, blind woodworkers—he replies in a calming southern drawl, "I'm sure my story will be secondary to all of those." But it's not.

Jimmy Carter isn't some craft show tinker cranking out wood-burned geese—he's a serious woodworker with a wide range of interests and a tremendous storehouse of skills.[19] He's made over a hundred pieces of furniture since leaving the White House—raised-panel armoires, natural-edge benches, pencil-post beds, church collection plates, ladderback chairs, green wood pitchforks, and whittled chest sets. He's made every lick of furniture for the small mountain cabin that he, and wife Rosalynn, built in northern Georgia. "I've made four different style cradles that I've given to my own children so they'd increase the production of grandchildren," he explains with a lilt to his voice.

Carter's earliest relationship with wood was based in practicalities. In a memoir of his childhood, he tells how one of his first chores was loading oak and hickory into the wood stove to start the morning fire. He helped his father maintain their tenants' wood cabins and helped plant a pecan tree orchard. "As I got older," he writes, "I helped with all the jobs in the shop, but was always most interested in working with wood, especially in shaping pieces with froe, plane, drawknife, and spoke shave."[20]

He learned the basics of woodworking from his father as well as from instructors while a member of the Future Farmers of America in high school. As a submarine officer in the Navy he continued woodworking, more out of economic necessity than enjoyment. "We saved money by renting unfurnished apartments, then I'd use the hobby shops the Navy provided for building furniture for them. So at first it was a practical matter," Carter explains. Some of his earlier pieces look clunky, as most beginning woodworkers' pieces might, but through these projects and under the guidance of the master carpenters manning the Navy shops, Carter was able to learn proper joinery, gluing techniques, and wood selection.

As president he made occasional use of the woodshop at Camp David to make small presents for friends, but it was only upon leaving office that his passion for woodworking ignited. "When I left the White House, my staff and members of the cabinet took up a collection to buy me a going-away present, and they were going to buy me a Jeep," he says. "But I found out about it, and didn't want it, and I told my secretary, who told them that what I really wanted was a completely outfitted workshop." They gave the $7,500 they'd collected to Sears and asked the company to furnish a complete shop. And that's what he got.

"Since I've never owned an automobile since then, I utilized our two-car garage as a workshop, and that's where I've maintained a complete shop of power tools and hand tools," Carter says. His lifetime Secret Service protection takes care of the driving.

While Carter owns and uses a full array of power tools, he's drawn to the woodworking practices of pioneers in the late eighteenth and early nineteenth centuries. Two of his favorite tools are the drawknife and the spoke shave. "I've particularly enjoyed using green wood," he says. "I go out behind my house, where I have a good stand of oak and hickory trees, and I use a handsaw—to emulate what people had to use years ago—and cut the tree. And my wife and I drag it to my woodshop, and then, using chisels, mallets, froes, a drawknife, and so forth, I make furniture out of the green wood." From these raw materials Carter has crafted entire sets of stools and greenwood chairs, complete with hand-woven hickory bark seats.

Carter explained in a prior interview that "there's a tremendous amount of work that comes with [working green wood]. You also have to develop an intimacy with the woodlot to start with—and then between you and a particular tree. You have to envision what your project is going to be like and then try to handle the different rates at which the wood dries, so the joints get tight and indestructible."[21] It's the kind of technical problem that appeals to the nuclear physicist in Carter.

Working with non–kiln-dried wood has also been one of his greatest sources of frustration. When I ask him about any notable woodworking blunders, he responds, "Well, generically speaking, the most mistakes I've made have revolved around shrinking wood after I've finished the piece. And I've learned from mistakes to be very careful about the curing of wood. I was too careless in some instances, and the piece came apart or had unresolved cracks in it." I sense that "unresolved cracks" is the politically correct term for firewood.

When I ask Carter if he had to select one piece of furniture he's made in which he takes the most pride, he hesitates. "That would be hard to say. I'm looking at one right in front of me in my office that I think would be among the top ones. We have a little mountain cabin in north Georgia, and while I was up there one day, one of my neighbors had a big chestnut stump that he'd dug up from his yard, and so I took part of that root and part of the trunk of the tree and I brought it home, and I just decided to use all hand tools and no power tools, and I made a coffee table out of it. It has rough edges, just the way the tree was shaped. And that's one of my favorites. I had to use some butterfly patches, since it was split in a few places, but I think that's one of my favorites."

In a follow-up question that I can chalk up only to irrational exuberance, I ask him if he doesn't think it's odd that while he cuts down trees and turns them into beautiful pieces of furniture, George Bush's favorite pastime, according to recent reports, is cutting down trees on his ranch and burning them. In the damp hollow recesses of my mind, I must be thinking I can garner a quote of monumental import—one that encapsulates the worlds of wood and political ideology and wraps them up into one single, immortal sound bite. And even as I'm mouthing these words, I realize I'm making a huge mistake. That perhaps the reason he's granted me this interview is precisely because it is nonpolitical. That perhaps he's carved time out of his intensely busy schedule because it offers a reprieve from the torrent of criticism he's been receiving over his recent book, *Palestine: Peace or Apartheid.* And he indeed responds with that response no interviewer ever wants to hear—the interminable, dreaded, thud of silence. Next question!

Being the former president of the United States may help Carter get a foot in the door in the quest for woodworking advice, adventures, and inspiration, but one gets a sense that even without the presidential seal embossed on his business card, Carter would ferret out the best woodworkers in his travels. He's been able to visit with legendary woodworkers like Sam Maloof and Tage Frid. "I've traveled in different parts of the world—in Japan, China, and so forth—and I've gone out of my way to meet some of the finest woodworkers on earth, to talk with them and get their advice on hand tools to acquire." One of his favorite side trips was visiting a woodworker in Ghana who builds custom-shaped caskets based on the interests of the deceased. "There are carrots, hot peppers, or ears of corn for farmers; boats, fish, squid, or different kinds of sea life for fishermen; and automobiles, buses, or airplanes if the customers have these special interests. His most interesting addition on our last visit was a jogging shoe, with carefully crafted tongue and strings, in which one of Ghana's Olympic runners plans to be interred."[22]

Carter applies what he's learned to good causes. He's made offering plates and other items for his church. For each of the past fourteen years, Carter has made a signed piece of furniture or other woodworking item to be auctioned off as a fund-raiser for the Carter Center at Emory University near Atlanta. Pieces have ranged from steam-bent rakes and pitchforks to elaborate cabinets. One of his recent pieces—a tall cabinet made of persimmon—sold for an even $1 million.

Carter has penned books on aging, faith, relationships, outdoor adventures, and, of course, politics. Soon he'll be adding a woodworking title to the list of his many published works: a coffee-table book with photographs of sixty-five pieces of furniture he's built, complete with descriptions of the woods, joints, and impetus for building each piece.

A conversation with Carter would be incomplete without a discussion of a different type of wood—the two-by-fours he bangs together in his work with Habitat for Humanity. He's been involved with the nonprofit group since 1984, when he helped renovate a six-story building in New York City to provide housing for nineteen families. Since then he's organized the annual Jimmy Carter Work Project, which has drawn tens of thousands of volunteers to build houses in such places as Mexico, India, South Korea, and a variety of American cities. Most are blitzes, when entire homes are completed in five-day stints.

"In America it takes about thirty-five volunteers to build a house, and they average about twelve hundred square feet," Carter says. "In foreign countries like the Philippines, South Africa, or India, the homes are between three hundred and five hundred square feet, and it takes twelve or fifteen people to finish a house in five days."

When they finish, Habitat volunteers often leave their tools behind so the new homeowners—who are required to invest at least five hundred hours of labor in the low-cost, interest-free homes they purchase—can continue to upgrade and expand their homes. Habitat as a whole has built more than two hundred thousand houses worldwide.

"Habitat has opened up unprecedented opportunities for me to cross the chasm that separates those of us who are free, safe, financially secure, well fed and housed, and influential enough to shape our own destiny from our neighbors who enjoy few if any of these advantages in life."[23] Carter is much more than a figurehead for Habitat. Based on what those who have worked around him say, you'd better keep out of his way while the sawdust flies.

When I ask him if he can still sink a 16d nail in three whacks—a feat he displayed several years back on the Jay Leno show—this eighty-three-year-old carpenter replies in the affirmative. "Obviously, it depends on the kind of wood. I would say in a common two-by-four pine stud, yes."

In the end, the part he enjoys most about woodworking is the act of woodworking. Still a country boy at heart, he often rises early, writes from 5 a.m. until 10 a.m., then heads to the woodshop to design and build furniture. "When I tire of the computer screen, I can walk twenty steps to my woodshop and immerse myself in my current project," Carter says.[24] And why does an octogenarian, financially secure, Nobel Peace Prize recipient, Emory University professor, Carter Center (Waging Peace. Fighting Disease. Building Hope.) leader, and prolific author continue to make sawdust?

"What we need in our lives is an inventory of factors that never change. I think that skill with one's own hands—whether it's tilling the soil, building a house, making a piece of furniture, playing a violin, or painting a painting—is something that doesn't change with the vicissitudes of life. [Woodworking is] a kind of therapy, but it's also a stabilizing force in my life—a total rest for my mind."[25]

Whether it's building a house for Habitat for Humanity or carving a wood Ferrari for cruising the canals of Venice, those who work with wood share one trait in common: tool lust. It may be lust for the latest tool, for the oldest tool, for the fastest tool, for the prettiest tool, or for the safest tool—but the lust is there.

It would take a Paul Bunyan–size awl to even scratch the surface when it comes to covering the field of tools. One book is dedicated solely to boxwood and ivory rulers made by Stanley; another to embossed American axes; and yet another to tools made by the Sandusky Tool Company in 1925. So while it's not possible to tell all the stories, it is possible to peek at a few of the more interesting ones.

CHAPTER 3

The Tools That Work the Wood

Walking through the Louvre in Paris last year, I found the *Mona Lisa* to be seriously overhyped and seriously underlit. Had it not been for the crowd gathered around the Rembrandt sketches, I would have strolled right by. And the *Venus de Milo*—at least to me—seemed to be missing a certain something. Like arms. But one work of art stopped me dead in my guided audio tour tracks.

It was the painting *St. Joseph in the Carpenter Shop,* by Georges de la Tour, from 1642. The only source of light in the painting is a single candle cradled in the hand of Jesus, then aged about seven. His left hand is positioned between the candle and the viewer, so the flame is concealed, but the light imparts a glow to select objects in the room. It illuminates the face of Jesus, fixed in a look of innocent admiration as he watches his father work. It highlights only the eyes, nose, and beard of the bent-over Joseph—yet within those sparse features you can detect the crestfallen look of someone who knows what's coming down the pike.

The candle also illuminates the tools Joseph is working with. He's turning an auger as he bores a hole in a large timber lying on the floor; a thin curl of wood nearby, a testament to the tool's sharpness. A mallet and a chisel rest beside him. The cross-shaped auger, cross-shaped mallet, and cross-shaped timbers have a story to tell; they foreshadow the inevitable.

Many tools have stories to tell. Here we'll look at a few of them: The stories behind lathes, belt sanders, table saws, and collectors.

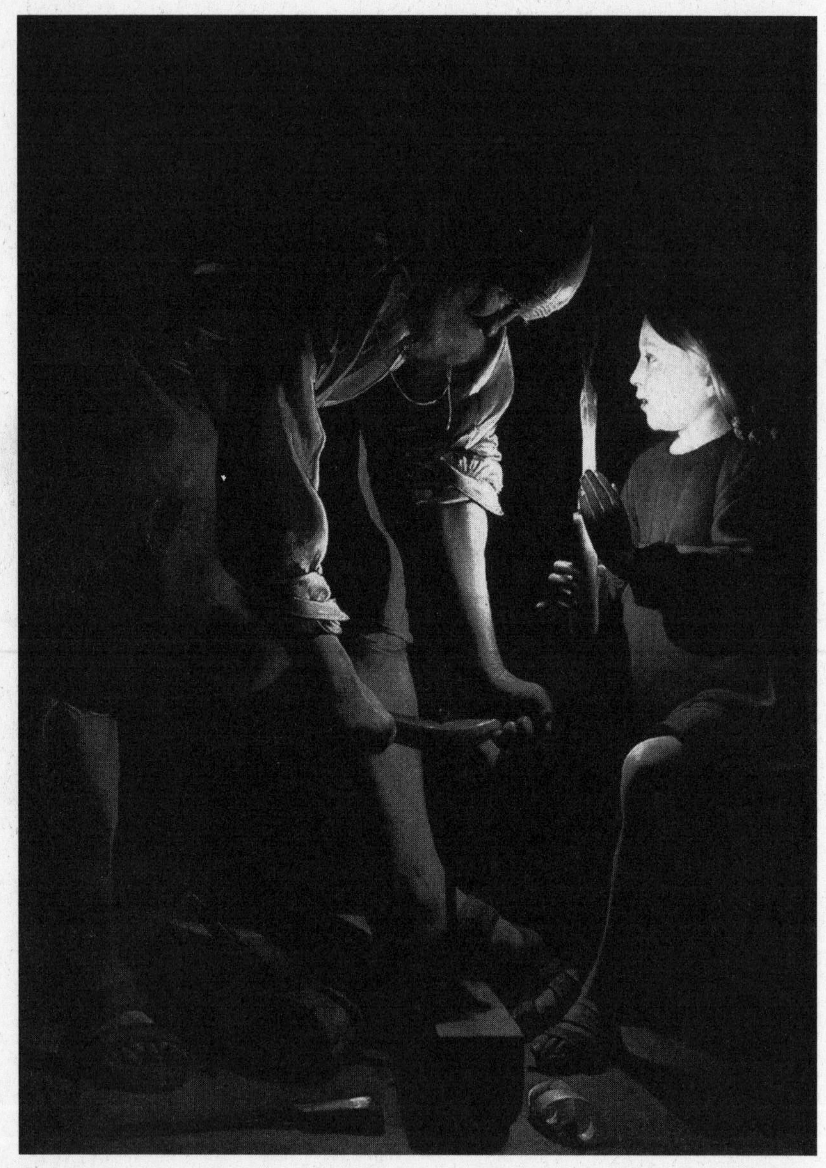

St. Joseph in the Carpenter Shop, painted by Georges de La Tour in the 1640s, now hangs in the Louvre in Paris. The wood and cross-shaped tools foreshadow things to come.

AS THE LATHE TURNS: MAKING GOLF TEES WITH THE MASTER

As I stand in Alan Lacer's workshop I feel like a seven-year-old T-ball player asking Ken Griffey, Jr., for a batting lesson. My turning equipment at home consists of four cheap gouges and a Ryobi minilathe so insubstantial I can carry it under one arm. My lifetime turning output consists of eight pens, five rolling pins, and five Christmas tree ornaments, the last of which doesn't count, since even the most botched turning project, with a string and hook attached, can be passed off as a Christmas tree ornament.

Lacer's skill level and tool arsenal are a quantum leap beyond mine. His turning studio (for whatever reason, carvers and wood turners have "studios," while cabinetmakers and furniture makers have "shops") is in a converted barn in rural Wisconsin and contains lathes of every conceivable shape and size. He has one lathe with a 12-foot-long bed and a belt-drive motor mounted on the ceiling that he uses for turning 8-foot-tall porch posts. Sitting next to it is a working lathe the size of a work boot. He has a cast-iron treadle lathe from the late 1800s sitting next to a Stubby lathe that runs as smoothly and quietly as an Audi, with price tag to match. There's a massive tracing lathe that—by copying parts placed in a special jig beside it—will spit out identical spindles or other turnings all day long.

I'm attending a day-long turning class along with four coworkers, and contrary to what I thought I would be walking out with at the end of the day—a nicely crafted table leg or tool handle—I walk out with an 8-inch-long golf tee. Based on that level of production, you might suppose I am a slow learner or that Lacer is a bad teacher, but at least on the last count, you'd be wrong. I am thrilled with my golf tee.

The day spins by. We spend one hour chatting about the history of lathes, another hour eating at the South Fork greasy spoon, and a half hour righting a refurbished bandsaw that's lain prone on Lacer's workshop floor for the past eight months. But fully half of the day is spent sharpening. "Turning should be effortless," Lacer explains.

"The tool should do the work." If you're fighting with a dull tool, everything—lathe, wood, tool, turner—becomes stressed and overworked. And though instinct might say a sharp tool is dangerous, it's the dull ones that catch an edge or are forced that cause the most injuries. "Lathes are incredibly safe," explains Lacer. "It's the only stationary power tool where you hold on to the cutter and you push the sharpened end away from you. I've worked with kids as young as three and a half years old on the lathe."

Safety is a big plus, but there are other equally strong attributes. Turning is quiet, requires few tools, and is relatively mess-free. Some apartment dwellers have been known to turn in their kitchens without complaints from neighbors or spouses.

Turning is also a fabulous creative outlet. Bowls, pedestals, spaghetti jars, vases, spindles, pens, goblets, and a hundred more items can spin off a lathe. You can turn bone, antlers, plastic, ivory, tonga nuts, ice, aluminum, slate, even phone books if they've been soaked in epoxy first. Lacer has even turned vegetables, explaining how in one class he was turning large rogue carrots while a student was catching shavings in his mouth.

Turning can also provide instant results. It's the only kind of woodworking where the raw material can have a bird perched on it in the morning and be used as a bread bowl that evening. It allows you to crank out a forgotten birthday present in a few minutes and still make it to the surprise party on time. And if the gift is a wine stopper, you'll probably be able to pick up orders for a few while you're there.

One of the great beauties of turning is its broad appeal. Turners include amateurs who like instant gratification, professionals who are raging perfectionists, and every mindset in between. Even Hollywood hotshot William Macy is an avid turner, having been bitten by the turning bug while filming *Fargo*.

Turned works have attained the status of fine art, where it's not only important *what* the thing looks like but also *who* made it. Professional turner Bill Hunter recently sold three pieces at the Sculptural Objects and Functional Art (SOFA) exposition in Chicago for over $20,000 each. An unsold piece had a $40,000 price tag. Johannes

Johannes Michelsen sporting one of his lathe-turned Range Rider hats. Prices range from $600 for a baseball cap up to $4,800 for his most intricate creation.

Michelsen of Vermont specializes in turning wooden hats. They weigh only 6 to 9 ounces, and you can actually wear them. For $1,350 you can own a Range Rider cowboy hat made of walnut burl; for a more casual look, consider the $950 baseball cap made of madrone wood. Malcolm Tibbetts creates segmented wooden sculptures consisting of turned shapes that have been cut up, then reconfigured into other shapes that may or may not even be circular. One recent work, Galactic Journey, consists of 12,960 individual pieces of wood.

Turners are the Hells Angels of the woodworking world. It's a craft where anything goes. Some in the established woodworking world will frown upon turners—but they have a lot of fun.

For most of the twentieth century, turning was done to make parts for bigger projects—table legs, bedposts, and stairway spindles. As an art form it was the province of few. But when the craft fairs of the 1960s and 1970s emerged, suddenly turnings weren't just parts of other things but things in and of themselves. Turners had a venue for getting their work out in front of lots of people. They got immediate feedback. They pocketed most of the money from the sales, so they could actually make a living. Craft shows gave woodturning the shot in the arm it needed. Now the craft thrives. The American Association of Woodturners alone claims thirteen thousand members.

Specialty manufacturers are creating lathes and jigs that make it ever easier and easier to create more complex pieces. Some turners now even use laser calipers for measuring and maintaining the thickness of their bowls and vessels as they work.

This mechanization and technology infiltrating some facets of turning lead Lacer to wonder if things haven't gone too far. He refers to a principle David Pye lays down in *The Nature and Art of Workmanship*. In the book, Pye examines the craftsmanship of certainty versus the craftsmanship of risk. Lacer holds up the store-bought coffee mug he's sipping coffee from to illustrate his point. "This is the craftsmanship of certainty," he explains. "You know what the final product is going to look like. Someone poured some clay into a mold, the cup comes out perfect, and that's it."

"But wood turning has always been about the craftsmanship of risk," Lacer continues. "Things aren't perfect, and that very imperfection is what makes a wood vase or bowl perfect. That's part of the warmth and beauty of it. That's the fun of it."

The craftsmanship of risk goes way back with the lathe. The earliest indications of the lathe's existence are turned artifacts from Europe—an eighth-century BC Etruscan bowl and a sixth-century BC Bavarian bowl. While the bowls survived, the lathes themselves did not. The earliest depiction of a lathe dates to the third century BC, where an Egyptian grave contains an image of one in bas-relief. It's depicted as being used vertically; most lathes eventually evolved to the horizontal position.[1]

Bow lathes emerged early on. These involved the turner winding the loose string of a bow around a cylinder or pole several times, then tensioning the bow and moving it back and forth in a handsawlike motion to create a spinning motion. Spring pole lathes emerged in the

Alan Lacer test-drives a replica of an early wooden treadle lathe. The tools of his trade date back to at least 800 BC.

1200s in Europe and remained in common use well into the twentieth century. These were essentially large bow lathes, with the power supplied to the string and drive train by the downward thrust of a pedal, followed by the upward thrust of a springy pole. The main drawback to these early lathes was that their rotation alternated every second or two between clockwise and counterclockwise; you could cut only half the time on the clockwise part of the rotation.

Eventually, lathes that spun in a single direction were developed, so turners could work full time. Wheel lathes operated like gigantic bicycle gears, with a large hand-powered wheel delivering power to a much smaller and faster rotating lathe wheel. Foot-operated treadle and velocipede lathes emerged next. "Lathes have

"The Juggler," by turner Malcolm Tibbetts, is made of curly maple and other exotic woods. The top sphere is a lidded container, which the artist calls a multigenerational lamination. Some of his segmented turnings contain over 12,000 pieces of wood.

been around a long time," says Lacer. "The appeal is primitive. Essentially, all that's been added over the centuries is a motor."

The basic techniques also haven't changed much. In our class, Lacer works with us on these principles as we turn our golf tees. Start with the bevel of the tool in line with the wood, then raise the handle and push to cut. Keep the part of the tool doing the cutting supported on the rest. Always cut downhill. Work in slow increments.

We learn to adjust the speed of the lathe based on the diameter of the piece being turned. The outer rim of a 16-inch bowl turning at 600 revolutions per minute is traveling at 29 miles per hour; the outer rim of a pen turning at that same 600 rpm is traveling at 1 mile per hour. Kick it up to 1,000 rpm, and your bowl rim is approaching 50 miles per hour. Lathe speed is not to be taken lightly. That's why the ultra-careful wear not just a face shield but goggles under the face shield.

We all start out with a 2 × 2 inch block of wood. The goal is to turn this into a perfect cylinder and then to keep turning this perfect cylinder into smaller and smaller perfect cylinders until a golf tee shape is attained and the piece snaps and falls harmlessly to the bed of the lathe.

Since we're turning golf tees, I think in golf analogies. Both activities involve hundreds of small nuances—ways of holding the club, selecting the right club, body position—that make the difference between a hole in one and a muffed shot. And in the same way Tiger Woods makes swinging a seven-iron look easy, Lacer makes wielding the skew chisel look effortless and second nature.

Lacer does it all. He turns transparent goblets, with 1/16-inch walls so thin that light shines through them. He turns dinky tops, some as small as 1/4 inch in height—and to add to the level of difficulty, he does it with a 1 3/8-inch roughing gouge. He turns tall-lidded Russian spaghetti boxes for storing pasta and thumb-size chili pepper canteens. Rough-edge bowls, spalted wood, porch posts—you name it. High on the difficulty scale are the flying saucer–shaped vases he turns. They have impossibly small openings and perfectly uniform wall thickness; making them involves blind turning, in which the turner can't see the tip of the gouge, and touch is everything. "I still lose about 40 percent of the bowls I turn this way," explains Lacer (some of them painfully near a state of completion). "But after a while you just accept that as part of doing work like this."

At the end of the day we walk over to Lacer's house, which, not so surprisingly, holds a compendium of turned objects from around the world. He has tops from Japan, a place he's visited numerous times to conduct turning classes. "Tops are huge in Japan," Lacer says. "They have over a thousand different types of tops over there." He pulls out wonderfully turned and painted kokeshi dolls, toys historically created by itinerant woodworkers who were too poor to buy conventional dolls for their children. He opens up a small box containing wood horses and tells us they're turned—an absolute impossibility—until Lacer shows us a piece of wood in the shape of a doughnut with a slice cut out. Looking at the exposed ends of the doughnut, you see the perfect profile of a horse that's been turned into this once-360-degree ring. The horses are sliced off like little pieces of doughnut until the doughnut is no more.

As we head out the door at the end of the day, Lacer gives each of us a transfer-type tattoo on a small piece of paper. It's a tattoo of a wood-turning chisel. It's the kind that will wash off in a day or two, but it's easy to see how turning could stay with you for life.[2]

TOOL JUNKY HEAVEN

The displays of rope twisters, corn shellers, and old-fashioned prostate surgical tools (ones I know I don't want anywhere near *my* prostate) attest to the fact that the word "tool" has a definition that far transcends the world of woodworking. But at the Mid-West Tool Collectors Association spring meeting, it's clear that woodworking tools rule.

Tom Caspar, senior editor of *American Woodworker* magazine and hand tool aficionado, has tried to prepare me ahead of time for who and what I may encounter. He forms circles with the thumb and pointer finger of either hand. "This is the world of woodworkers," he explains, nodding toward his left hand. "And this is the world of tool collectors," he explains, nodding to his right. Overlapping the circles just a sliver, he says, "The two worlds really don't overlap that much. To become adept at using hand tools well takes a lot of time and practice. It's a hurdle a lot of people don't make. But to be a collector, well, all you have to do is collect things. People collect all sorts of bizarre things," he adds.

And I do find a lot of bizarre things. There's the Whatzit? table, where members place mysterious-looking tools and other members attempt to guess their use. Eventually I learn the true identities of the mystery tools. Item 125-2, which resembles a bugle run over by a freight train, is actually a pruning saw. Another item, a handsaw with handles on both ends, turns out to be a wedding saw that is described in this manner:

> The people in the village would set up a sawhorse, a log, and a double-handled saw. The newlyweds must saw the log apart with the prompting and cheering of the crowd. When the job is finished and the log cut, it symbolized that the man and woman must work together in all of life's tasks.[3]

This is surely the most romantic tool in the entire building. Overall, it's an orgy of tools for sale, for trade, for display, and for conversation. As with any group of serious collectors, minutiae and specialization reign. One member specializes in nail pullers and has written a 416-page book about them. Another has over two thousand hammers in his collection. Yet another collects 4-inch pocket rules exclusively. One display exhibits nothing but No. 2 planes—sixty of them—along with a sign containing an insider joke that reads "No. 2 alike." A couple who has come all the way from England sells brass plumb bobs the size of grapefruit crafted by a retired Rolls Royce machinist.

Harold Barker of Ada, Ohio, has one of the more unique booths. He's spent thirty-five years compiling catalogs, journals, and newspaper clippings into pictorial history books that document the development of, well, nearly everything. Amid his hundred-plus books—each

The "Big Johnny" double-handled sledgehammer required two men to swing and was used for whacking railroad ties into place. This is just one of over two thousand hammers in Scotty Fulton's hammer collection.

hand printed using a photocopier—you'll find a 148-page *Pictorial Survey of American Blow Torches from 1890 to 1950,* two volumes on ice harvesting tools, a 75-page book on brooms, and five pictorial histories of the lathe.

The plethora of hand planes drives home the point that before the advent of power tools, these tools served as router, surface planer, jointer, belt sander, and shaper. Some are as common as hair, with price tags of $10 to $25 to match. But if you're looking for the highest-priced tools, you look to the plow plane. Tom Witte, who's been in the tool selling business for thirty years, has sold one for $20,000 but knows of another that recently sold at auction for $140,000.

I look at a solid brass tool that's labeled "powder room brace" and, presuming it has something to do with bathrooms, ask Tom about the special use of this early drill. "Whenever they'd work in the rooms where they stored gunpowder, they'd use brass tools, so if they dropped it, it wouldn't spark." He has other unique braces: an undertaker's brace, which held a screwdriver bit and could fold up into his back pocket, so when it was time to close and seal the coffin, he could do it quickly, then put the tool out of sight. And he has braces with handles of horn, ebony, and rosewood, some with price tags in the $2,000 range.

A few tables down, I run into another strange brace—one that has no place for a bit—and ask the owner what it is. "I was hoping someone would ask," he howls, picking it up and pressing the tip against my back. He starts cranking the handle, and as I feel a soothing massage I realize it's a hand-cranked vibrator. "A few years back at one of these meets my wife had a sore thigh and asked me to use this on her. We were in our motel room, and I was cranking away when a buddy walks in. He thinks I'm using it on a different part of her anatomy. Oh, we laugh about that one."

Despite its small-scale regional-sounding moniker, the MWTCA is a large-scale international organization with more than three thousand members, who hail from Italy, France, England, Manitoba, Australia, and Norway. It's such a gray-haired crowd that the registration form has a box to check if you want to reserve an electric handicap scooter. At one point I'm looking for the head of the meet; the people at the information table tell me to look for a balding guy in suspenders, and when I look around the exhibit hall realize that doesn't narrow the field much. There's a tour of the Spam Museum in Austin a few miles away for the ladies' auxiliary, and Friday afternoon there's a presentation on "hardware and axe brands and markings common to Minnesota, Wisconsin, and Iowa."

Before I leave, I purchase a sixty-five-year-old brass plumb bob as a going-away gift for a coworker. It's old but shiny, once hard working but now retired, an entity with lots of stories to tell. Nice. A little bit like those around me.

THE TABLE SAW THAT COULDN'T CUT A HOT DOG IN HALF

In a scene that only a fervent woodworker could appreciate, fifteen grown men stand huddled around a table saw, watching a one-by-six board inch slowly toward the blade whirring at 4,000 rpm. Everything appears normal except for the Hormel hot dog sitting atop the board. We all watch in anticipation to see if this 5-horsepower 640-pound machine can cut this gastronomic American icon in half. As the blade contacts the hot dog there is a violent explosion. The blade stops in milliseconds, then dissolves into thin air. The operator, a woodworker by the name of Richard Tendick, holds up the hot dog, unscathed save for a 1/32-inch nick. A small chorus of oohs and ahhs ensues. The table saw has performed exactly as intended.

Tendick removes the throat plate from the surface of the table saw, and fifteen tool fanatics crane their necks in unison to stare into the working—or at least once-working—mechanism. The blade is toast; a fist-sized brake has jammed an aluminum block into the teeth of the blade, bringing it to a halt in 1/200 second. The $75 brake cartridge must also be replaced, but the Hormel surrogate finger has shown that the SawStop is one hell of an injury preventer.

A table saw equipped with the woodworking equivalent of an

The SawStop about to do its thing as another test hot dog approaches the blade. The blade will be brought to a stop in less than 1/200 second once the hot dog contacts a tooth.

air bag may seem like overkill until you look at the statistics. Table saw injuries send over thirty thousand people annually to the emergency room, and three thousand of those people exit with the inability to count past nine on their fingers. If you ask Benjamin O. Powell, who lost his pinky finger and 40 percent of the palm of his hand in a kickback accident, he'll tell you he would have walked away with a nick instead of an amputation that cost him two months' work and $50,000 in medical costs if he'd had the device. The $1,000 extra cost of the SawStop table saw looks dirt cheap in comparison.[4] Or chat with James Kissick, who teaches construction trades at Swift High School. One of his students who recently plowed his hand into a SawStop blade incurred not an amputation but a nick so minor it was closed with Krazy Glue.

Saws have been around for at least fifteen thousand years, perhaps as long as a hundred thousand. The earliest were "found" objects: the jawbone of a sharp-toothed snake or the backbone of a spiny fish. The first manmade versions consisted of rows of flint, obsidian, sharks' teeth, or seashells embedded in wood. Bronze saws eventually emerged, but they dulled quickly. It wasn't until the art of ironmaking developed that saws became truly efficient tools. Many early iron saws consisted of a blade held in tension, operated by two people. The larger versions of these saws eventually evolved into pit saws—muscle-intensive operations involving two sawyers, one standing above the log pulling up, the other below in a pit pulling down. These remained the tool of choice for turning trees into boards for centuries, and they are still used in some developing countries.

Though mechanization eventually took over, a fellow with the last name of Disston kept the handsaw industry alive and cutting with his specialty saws. A list from the early days reveals tools specializing in nearly everything: ice saws, lock saws, meat saws, chair web saws, sash saws, stave sawing machines, washing-tub saws, and subcutaneous saws. For more information you can visit the Disstonian Institute on the Internet.

Water wheels were harnessed to power saws as early as the fourth century in Germany and were surely operating in France and Norway in the fourteenth and fifteenth centuries.[5] But as logical as it may seem to us today, the circular blade—the shape that could make the most efficient use of a water wheel's circular motion—wasn't invented until the 1770s (though some claim the Dutch beat this by a hundred years). Until the development of the round blade, the circular motion of the water wheel was converted into a reciprocating, back-and-forth, or up-and-down motion mimicking that of a handsaw—not particularly efficient.

Blade design wasn't the only problem pestering sawmill owners of that early era. From the 1500s right up through the 1800s, it was common for mobs of hand sawyers to torch new-fangled mills to ensure job security. In 1802, an enterprising fellow from Philadelphia harnessed the steam engine from a grounded riverboat and was able to saw a then-phenomenal 3,000 board feet a day—until out-of-work hand sawyers burned his contraption.[6] But circular saws were not to be denied, for they were efficient machines indeed. Initially, the blades were large clumsy affairs hammered out by blacksmiths. But as metals, designs, and sharpening techniques improved, so did the efficiency of the sawmills.

By 1852, *Scientific American* was publishing sketches of machines that roughly resembled the table saws of today; some with tilting tables for creating beveled cuts. One might surmise that the stock of artificial limb manufacturers rose dramatically in the mid-1800s; not only were the blades fully exposed, but the pulleys, gears, flywheel, and belts that drove them were as well.

It didn't take long for the circular saw to hit full stride. In 1859, the Kaefer Power Company began manufacturing a combination tool that consisted of a circular saw, scroll saw, borer, and mortising machine "each arranged so as to be readily removed out of the way, and all worked by one treadle, in combination with a peculiar flywheel."[7] The same year, radial arm saws, gang saws, and self-feeding saws emerged.

Large-scale circular saws were originally dependent on water power, limiting their location to rivers or dams. When steam power came into play, sawmills could be built anywhere and on a smaller scale. Yet, well into the 1900s, manually operated treadle and hand-cranked saws remained popular for small operations. The Barnes Tool Company ads of the day touted their "E.F.B. Hand and Foot Power Sawing Machine" with quotes from their customers: "My machine is worth four of my best men" and "It is worth $1.00 per day to any small shop, to say nothing of a large one."

The electric motor finally brought table saws down in size and cost, so even a small-scale woodworker could accommodate one. The 1925 Beaver Woodworker could "operate from any electric light socket" and could be had for only $35 down. The diminutive 6-inch Delta circular saw, which one could bolt to the workbench, cost a mere $42 and weighed a mere 90 pounds. By 1932, Driver Power Tools was promoting a new feature of their saws with headlines blaring "TILT THE SAW—NOT THE TABLE!"

The first sighting of a table saw safety device didn't occur until the turn of the century. In 1902, the J. A. Fay and Egan Company offered a table saw with their Patent Saw Guard "at slight extra cost." According to their catalog, "The laws of many countries require the machine be equipped with a saw guard, and its possession prevents injury to employees and damage suits which attend injury where saw guards have not been provided." Even the cheapest table saw today comes equipped with a unibody tripurpose guard, which includes a shroud for covering the blade, a splitter for keeping the cut spread to prevent blade binding, and antikickback pawls for biting into the board to minimize the chance of the blade hurling the board back toward the operator.

But clearly, these safeguards are not enough, as witnessed by those thirty thousand visits to the emergency room per year. When you talk to those who have experienced table saw accidents, you nearly always hear the phrase "so fast I didn't know what hit me" or "my last cut of the day" or "wasn't paying attention for just a second." And that's where SawStop and Stephen Gass, its inventor, enter in.

Gass comes from a long line of hobbyist woodworkers and has used a table saw "ever since [I] could look over the top." He became intrigued by the notion of creating a device that could stop a blade before it could inflict harm. With a doctorate in physics, he ran the calculations, which included such ethereal considerations as how deep a cut could be before it crossed the line between minor injury and severe damage; a cut less than 1/8 inch deep became the goal. Then, based on the average rate at which wood (and therefore fingers) were fed into the blade, he determined how quickly a blade would need to be brought to a stop to meet his goal. He bought a used Craftsman table saw and started experimenting, working from the notion that different materials conduct electrical charges, with varying degrees of success. He developed a device in which an electrical charge running through the blade could sense the "large inherent electrical capacitance and conductivity" of the human body (or, for demonstration purposes, its wienerlike equivalent) and trigger the blade brake when it contacted skin.

Like all good inventors, such as Pierre Curie, who strapped radium to his own arm to determine whether it would burn the skin, Gass used himself as the guinea pig in the final testing of the tool. He was ready to ship his first saws off to a trade show, and he just *knew* people would ask him if he'd performed the ultimate test.

"I thought a lot about which finger to use for the test," says Gass. "I decided on the left hand because I'm right-handed and then decided on the ring finger because it seemed like the most dispensable." Then he gritted his teeth and tried feeding his finger through. The device worked, and Gass has not only all ten fingers but a bourgeoning business to boot.

There is a movement afoot—driven in part by the SawStop folks—to make blade brakes mandatory on all new table saws. "The numbers tell the story," says Gass. The average selling price of a table saw is $300, but the average societal costs from injuries, including hospital bills and lost wages, averages out to be $2,600 to $3,100 per saw.

Whether or not the legislation passes, I know one thing for certain: SawStop is clearly a better safety device than the one used in my junior high school shop class many years ago. That safety device consisted of the tip of a former student's finger, preserved in a container of formaldehyde dangling next to the table saw's ON-OFF switch.

BELT SANDER RACING: A SAGA OF TRUE GRIT, SPEED, AND VICTORY (SORT OF)

I'm standing amid a throng of tool geeks and gear heads near the starting line of the belt sander drag strip, in the Glarna Stuba bar at Tyrol Basin ski area. I'm not just standing amid the geeks and gear heads; I have become one. I've spent five evenings souping up my aged Ryobi 3 × 21 inch belt sander. I've switched two gears around to double its 1,250 SFPM (sanding feet per minute) speed, screwed furniture casters to the sides so it will glide off the track's side rails rather than bash into them, shortened and reglued the drive belt, and most important, mounted a Lucky Troll doll to the front as a hood ornament.

Fortunately I'm entered in the stock class, because compared with the heavy metal monsters entered in the open modified class, my Troll-Dozer is a castrato. In the stock class you can fiddle around

Author setting up his Troll-Dozer at the starting line of the Tyrol Basin World Championship Belt Sander Races. His second-place finish earned him a hearty pat on the back.

with the machines a little to goose the speed, but for the most part they still look like woodworking tools. But the only rules of the open modified class are that the machines be belt driven, under 20 amps, and less than 36 inches in length. There are machines made out of pairs of high-speed angle grinders, mammoth electric motors, and other drive devices unknown because the builders have disguised the inner workings. Bicycle chains and sprockets have replaced flimsy rubber drive belts and toothed gears. Some of the machines can hit

6,500 rpm. One of the entrants confides that he's spent over $1,000 and 125 hours building his machine, which looks more like a World War II German tank than a power tool. Some have belts 50 inches in circumference and are nearly as wide as the 8-inch-wide track. If one were to attempt sanding wood with any of these, they could halve a two-by-ten in seconds.

These tools are a long way from the tool a young Porter-Cable engineer by the name of Art Emmons invented eighty years ago. When Emmons invented the Take-About-Sander in 1926, he did so with the idea that workers would prefer bringing a hand-held tool to a bulky work piece rather than bringing a bulky work piece to a large stationary machine. And he was right. Though his hand-held tool—soon to become known as the belt sander—cost more than a month's wages for a tradesman of the day, it caught on quickly. Boatbuilders, cabinetmakers, furniture makers, and carpenters soon found the tool indispensable. Little did Emmons realize that a few decades later the tool would also bring me fame and fortune—well, at least fame; well, at least fun—at the Sixth Annual World Championship Belt Sander Races.

The entrants in the stock class have mild-mannered monikers like Sparky, Putt-Putt, Non-Big Racer, and my own Troll-Dozer. But those in the open modified class carry steroidal names such as Maniac, Tank, Rat Racer, and Twin Blades. The drivers are equally intense. Most are from the area. Many are machinists or mechanics, know one another, and are return competitors. The race is called the world championship because, according to Josh the bartender, who organizes the event, "no one else is calling their race the world championship, so we're calling ours that."

It's the first belt sander race I've attended, and a few things take me by surprise. First, I had imagined the start would take place by a flick of the switch that would start both machines simultaneously to propel them down the 40-foot-long track. But instead, it's more like a real drag race—one that brings the human element of reaction time and reflexes into play. There is a "Christmas tree" stand of starting lights that blinks down until the green light comes on. Only then can you slam down your big red button to start your machine. Hit your button too soon and you foul, automatically losing that heat.

I'm also surprised by the sophistication of the timing device. It's computer driven and accurate to 1/100 second. It tells you not just the speed of your sander but the speed of your reaction time from the time the green light flashes to the time you hit your START button.

And I'm surprised at the action that takes place *after* the machines cross the finish line. There's only 16 feet of "slow down" track, and, lacking parachutes or brakes, many of the faster machines come to a halt only by smashing into the wall at the end. It's brutal. The machines need to be not only fast but durable. Lots of things break—drive belts, gears, cowls, the wall—but not one sanding belt bites the dust permanently. These are devices with true grit.

The concept of sandpaper has been with us for a long time. In the thirteenth century, the Chinese began using natural gum to glue sand and crushed shells to parchment paper to create an early form of sandpaper. Hawaiians used granulated coral long ago to smooth surfboards. Early cabinet and furniture makers were known to have used dried sharkskin for sandpaper.

The first true sandpapers, actually made of sand and paper, were being sold in Paris as early as 1769.[8] The earliest sandpaper-related patent was issued in 1835 for a machine that manufactured the product. Most sandpapers today contain no sand, and very few belt sander belts even contain paper. The "sand" is most often flint, garnet, aluminum oxide, and silicon carbide. The "paper" is usually a woven material like rayon, polyester, film, or cotton. The belt on the Troll-Dozer is a 3 x 21 inch, 50 grit, 3M Regalite Resin Bond belt with Cubitron Abrasive and Y-weight backing—an impressive-sounding name and, because of its patented purple color, sharp looking too.

Hand-held belt sanders, as conceived by Emmons in the 1920s, are stock in trade for most carpenters and woodworkers today. And most owners have inadvertently raced theirs at least once by plugging in the machine with the trigger lock accidentally set in the ON position—then watching it propel itself off the workbench or across the raised cabinet door they've been painstakingly working on. Hand-held belt sanders top out at about 4 inches by 24 inches (4-inch width, 24-inch circumference), but there are sanders of a much larger scale. Benchtop belt sanders usually sport a belt in the 6-inch by 48-inch range. When you step up into the industrial-strength world of stationary belt sanders, you'll find machines running 52-inch by 103-inch belts. Some of these machines accommodate three belts of progressively finer grit, enabling you to feed a rough-sawn slab of lumber into one end and catch a finely sanded dining room tabletop on the other. No one would try to drag race one of these machines.

Well, there actually are a few who might give it a shot: the no-holds-barred bad-boy power tool drag racers. They stage their annual event at the Ace Automobile Dismantling scrap yard in San Francisco, and nearly any power tool is fair game: grinders, chainsaws, weed whackers, and circular saws are all allowed.[9]

There are a variety of classes. Superstock includes single-engine power tools driven primarily by belts or blades. The fine print in this category reads "No vehicles built from RC [radio-controlled] cars because they are boring." The Pro-supercharged class is similar to Superstock except that more than one motor is allowed. The Awful, Awful Altereds class allows all motors and power sources, including 208 volts single and three phase. The Unusual Designs/Top Fuel class is basically "anything goes," including "a double-barrel shotgun on roller skates pointed backwards down the track and fire[d] into a trash can filled with pig fat," as described in the rules. The fine print in this category reads "The tool/vehicle ITSELF must travel down the track (i.e. you can't fire a nail gun down the track and claim the nail is the vehicle)."

There is even a Ridden category with two classes: Funny Car and Sex Toy, the last of which is defined as "anything that can reasonably be claimed as a HAND-HELD sex toy [that] can be run (stock or modified), ridden or unridden. Use your imagination. But please, be careful. . . ." There are not only purses averaging $500 for first place per class, but also special prizes, including Most Pathetic Engineering, Most Spectacular Crash, and the much-coveted and monumentally politically incorrect Machine Most Likely to Get Its Maker Laid.

Chain driven, blade driven—anything goes at the Powertool Drag Races.

The Tyrol Basin belt sander races aren't nearly as frenetic, but they are fun. Snowboarders, beer, the clomp of ski boots, sawdust, and burning rubber all melt into one. It's so engaging that Madigan, the thirteen-year-old daughter of my friend and pit crew, Tom, has stopped text messaging for over an hour during the trial runs. Machines break apart and are patched together with gaffing tape, bolts, and superglue. There's a buzz as one of the machines in the open modified class runs a 1.92-second race—the first sub-2-second time in this competition's history, and a speed of over 20 miles per hour. Tracking, which involves keeping the belt running straight and true on the rollers, is a large issue. One can see that safety is paramount as Josh, the emcee, explains the rule that any machine jumping the track and landing on the floor gets beaten to death with a baseball bat by the Track Man so it can't do *that* again.

The stock class in which I'm entered starts with eight contestants. Two of the sanders break drive belts at the starting line, and two more are turtlesque, which narrows the field quickly to four, setting up the final bracket. I've snapped a belt, too, but I have had the amazing foresight to bring a spare. It's double elimination. I lose the first race but blow the dust bag off my opponent in the next heat with a run of 3.12 seconds. In the third race, my opponent fouls at the start line. And suddenly, as if in a dream, I'm in the World Championship finals. You could cut the tension with dental floss. My opponent is a guy named Sparky with lightning-fast reflexes, commandeering a $39 Taiwanese belt sander. I've been chatting with him in the bar beforehand and note that he's drinking lots of what looks like gin and tonic, which I hope will slow his response time. I, on the other hand, have been slamming Pepsi.

We position our machines and assume our positions in front of the start buttons. We both hit our "ready" indicator buttons, and the Christmas tree lights begin their countdown. The green light flashes, and I hesitate, slamming the start button a fraction of a second late: a fraction that costs precious feet. Sparky—who, as it turns out, has been pounding gen-X rocket fuel, Mountain Dew, not gin and tonic—is out of the gate first. But the Troll Dozer—clearly the machine with better aerodynamics, despite the big-haired doll screwed to the front—gains. The machines growl and roar as each senses victory. It's over in 3.02 seconds.

Alas, there can be but one victor. I emerge from the fray in second place. My trophy consists of a pat on the back from Josh and a smile. But it's not over. There's always next year. I want to do Art Emmons proud.[10]

When you look at the world of wood through the haze and sawdust of a belt sander race, it seems to be a rough-and-tumble blue-collar world. But throughout history, wood has shown a much gentler side—and perhaps the gentlest side of all is the one that faces the world of music.

It's a long journey from the belt sander races at Mt. Horeb, Wisconsin, to a violinmaker's workshop in Cremona, Italy, but it's worth the trip. Let's go.

CHAPTER 4

Wood in the World of Music

If you look at the value of things on an ounce-by-ounce basis, you find some very large numbers. According to the precious metal index today, you would find gold hovering near $1,000 an ounce, platinum at $1,756, and rhodium at a whopping $7,260. You would find the unusual, too. A cracker from Ernest Shackleton's unsuccessful Antarctic voyage—signed, of course—goes for $7,800, translating into over $50,000 per ounce. And a six-year-old brood mare by the name of Cash Run sold for $7.1 million in 2003, putting her value at $443 per ounce.

But in May 2006, a 17 1/2-ounce chunk of wood sold for $3.54 million, putting its value at over $202,000 per ounce. Of course it helps that this particular chunk of wood, known as the Hammer violin, was crafted by a man with the last name of Stradivari.[1]

Mystery, history, and brushes with fame play a large role in the value of a musical instrument. But no instrument can achieve fame unless it can deliver the goods in a serious way. And a good place to start looking for this quality is in the town of Cremona, Italy, the home and workshop of Antonio Stradivari.

STRADIVARIUS VIOLINS: THE SWEETEST SOUND YOU'VE NEVER HEARD

The curator of Cremona's Civica Collezione Di Violini stands in front of us, lecturing in Italian. Unfortunately, after twelve days in Italy, my vocabulary consists of *"Il conto, per favore"* (the check please) and *"pense che ci siamo perduti"* (I think we're lost). But whatever he is saying, he has the rapt attention of those around us.

He holds in his hands a violin made a few blocks from where we now sit: an instrument originally selling for a handful of florins in 1715 but today conservatively valued at $5 million. After five minutes of speaking, the curator—not inconsequentially, a master violinist—nestles the violin under his chin and draws, not tenderly, the bow across the strings, producing a sound that fills every crevice of the large room. It's a sound both brilliant and dark; an operatic voice that blends the highs and the lows, the joyful and the mysterious. And in that instant, I realize that in the hyped world of music, the sound of a Stradivarius is not hype. And I understand a little more Italian: *bella* (beautiful), *bravissimo* (excellent), *giocare su* (play on).

Three elements determine a stringed instrument's sound quality: the woods, the craftsmanship, and the design. Perfection in all three areas is essential, but which is *most* essential is a matter of an astonishing amount of research and scholarly debate. Later on the day of the concert, I go searching for answers at Cremona's International School for Violin Making. Giorgio Scolari, violinmaker and maestro; physics teacher Dr. Jorio Andrea; and student/interpreter Jonathon Hai have no lack of opinions in the matter. But in the attempt to steer our discussion toward the topic most concerned with here—wood—it soon becomes apparent it's impossible to discuss one element apart from the other two.[2]

Of all the parts of a violin, the top, or soundboard, is—albeit arguably—the most critical in terms of sound quality. It's Mother Nature's speaker cone; the part that both creates and transmits sound. The preferred wood today, and the same wood Stradivari used in his instruments three hundred years ago, is that of the red spruce growing in the Val di Fiemme area of Italy. The best wood comes from elevations of 3,200 feet or more: a cold environment that ensures trees grow slow, straight, and tall, which in turn produces straight-grained wood with tight, consistent growth rings. For stringed instrument soundboards, the ideal growth rings are consistent in size and spaced 1 to 1.5 millimeters apart. One supplier, the Ciresi Firm, cuts trees only in late autumn while the moon is waning. The soil is rich in silica, which is said to be absorbed by the tree, creating a wood with small particles of glass, which reflect light and sound beautifully.

The soundboard for each stringed instrument made at the International School is cut and prepared in the traditional manner. Sections of the spruce log a few inches longer than the instrument's body are quartersawn—a process equivalent to cutting a 2-foot-thick

A student, a maestro, and a physics professor (far right) show how and explain why nearly all of the instruments made by early Cremonese violinmakers were altered to accommodate the larger music halls and changing musical tastes of the day.

apple pie into slices. Each slice is then bandsawn once again into two smaller slices. These two pieces are then glued to each other "crust-to-crust" to create a wood blank that's symmetrical side-to-side, flat on the bottom, and about 1 inch thick in the middle, tapering to 1/4 inch at the two long edges. A blank of this shape allows the violinmaker to carve the arched top from solid wood. With the use of templates, calipers, chisels, planes, and infinite patience, the soundboard is coaxed into its final shape and thickness.

Purfling—a thin band of different colored woods—is inlaid 1/4 inch from the perimeter of the soundboard. Jonathon explains that the purfling serves two purposes: it's decorative, and it creates an ever-so-thin but solid band of wood along the fragile edges to inhibit cracking. But Dr. Andrea chimes in. He explains that these are not the most important tasks. The deep channel cut for the purfling and the purfling itself help isolate the body from the top so the soundboard can vibrate and resonate more freely: "Like a drrrum."

For the other parts of the violin, the species of wood are also carefully delegated for the jobs they have to perform; and within those confines, the exact pieces for each instrument are selected for their beauty. The traditional material for the neck and sides is flame maple; for the back, tiger stripe or fiddleback maple; for the fretboard, ebony. The woods of some violins are so spectacular that stringed instrument aficionados can identify the instrument simply by its back. The Il Cremonese violin, built by Stradivari in 1715, has a back made from a single board that may be the most stunning example of tiger stripe maple ever conjured up by nature or crafted by man.

The bow that's drawn across the strings to set the music in motion takes a backseat to nothing. Crafting one is an occupation in and of itself; rare is the violinmaker who makes the bow that plays his or her instrument. Snakewood—a wood that one bow maker described as ferociously expensive—is used for some bows, and ebony, with its near-immortal hardness, is often used for crafting the frog, which grasps the bow hair. But the ideal wood is pernambuco, or pau brasil: a wood of sublime weight, strength, responsiveness, and elasticity, which was once so abundant in Brazil that the country bears its namesake. Today it's one of the most expensive, and rapidly disappearing, woods in the world—and though a bow is slender, making one requires an inordinate amount of wood, a 2-ounce bow begins with a piece of wood fifteen times that weight.

Even selecting the 150 hairs that stretch across the bow is an art steeped in tradition. The masters prefer hair from the tails of horses that dwell in Mongolia or Serbia. And only stallion hair will do; mare hair—because of the direction of discharge—is too likely to be covered with urine, which can cause the hair to deteriorate or corrode faster.[3]

Most violins made today are based on the Cremona school of stringed instrument building, a technique that began with Andrea Amati's work on the violin's basic design in the 1560s, was refined by generations of other greats during the golden age of violin making—Gennaro, the Ruggeris, the Guarneris—and then culminated in the unequaled violins of Stradivari. Earl Carlyss, a former member of the Juilliard String Quartet, explains, "Strad was the one who finished the instrument. It has never changed since Strad's time—it ended with him. You couldn't improve on what he did."[4] This makes Stradivari's violins the most copied in the world, but herein lie conundrum No. 1 and conundrum No. 2.

Conundrum No. 1 is that there is no single Stradivari design. He was an experimenter and a scientist. He varied the lengths of violins, often by lengths best measured in millimeters, but enough to make a difference. He experimented with f-hole shape and position, with varnishes, and with woods and wood thicknesses. A visit to the section of the Museo Civico that displays the templates, molds, and tools Stradivari used bears this out. You'll find dozens of different body forms, paper templates for a variety of necks and f-holes, templates for checking contours, scrapers and planes the size of your thumb.

Conundrum No. 2 is that no one living today knows what an original Stradivarius truly sounds like. With but one exception—the

Medici Tenor viola, an instrument that stopped being used in the late eighteenth century because of changing tastes—every Stradivarius violin in existence has been altered, many dramatically.[5] As concert halls grew, as strings improved, as composers evolved, and as musical styles and tastes were modified, so were the instruments.

Dr. Andrea is adamant on this point. With wiry gray hair and wide eyes, looking every bit the part of the mad scientist, he points to the scroll—the small area containing the violin's tuning pegs—and explodes in a measured manner, with a heavy accent: "This is the only part unchanged. The angle and length of neck—changed! The height of the bridge—changed! The fingerboard and tailpiece—changed! The inside of the body—patches, patches, patches and changed! The finish—changed! The bow and strings—changed. Nothing's the same! Everything's different. No one knows what an original Stradivarius sounds like, yet everyone copies, copies, copies!"

Altered or not, there is an unmistakable quality to a Stradivari violin. Cecily Ward of the Cypress String Quartet plays a 1681 Strad, nicknamed the Fleming (based on its once-upon-a-time ownership by a relative of Ian Fleming). She explains that it's the immediate responsiveness and utter truthfulness of a Strad that makes it a Strad. The smallest nuances of the fingers—right or wrong—are translated into sound. Nothing is lost. The sound can fill a two-thousand-seat concert hall or an 8-by-8-foot practice room with equal grace. But why is it that while Stradivari's instruments are so often copied, the sound has not been?

Perhaps it's the wood? During the time Stradivarius was building violins, the wood supply was stringently controlled by government authorities. Trees were felled and floated downriver to Venice, where the best woods were set aside first for the vibrant shipbuilding industry. Woodworkers and wood merchants often had access only to what was left; wood sat in the swampy lagoons of Venice for months before being used. The lumber became imbued not only with fungi and bacteria but with minerals such as calcium, magnesium, sodium, copper, and iron.[6] These minerals affected the tone. Joseph Nagyvary, a biochemist at Texas A&M University, who has studied and attempted to replicate Stradivari's instruments in *exact* detail, believes that it wasn't the lagoons but the violinmakers themselves that marinated the wood in special blends. Some hypothesize that the woods were purposefully soaked for years.

Perhaps it's the varnish? Giacomo Stradivari, a great-great-grandson of the master, allegedly copied the formula he found written in the family Bible, which he then burned. He steadfastly refused to give up the formula on the grounds that "if by chance other Stradivaris—my sons, nephews, grandsons, or grand nephews—should turn their attention to the craft of our celebrated ancestor, they should then at least have the advantage of possessing the recipe of his varnish, the possession of which could not but be of material assistance to them."[7] Or "of material assistance" to Giacomo, since he attempted, unsuccessfully, to trade the recipe to a rare musical instrument dealer in exchange for a violin made by his great-great-grandpapa. Samples of Stradivari's varnish have been chemically analyzed, but to no avail. Older varnishes were a strange concoction of gums and resins bled from trees, oils from crushed fruits or seeds, solvents ranging from turpentine to alcohol, rosin and coloring agents derived from insects, plants, or mineral ores. He may have pretreated the raw wood with beeswax. There are too many variables.

Perhaps it was the weather? A series of cold spells, one starting in 1650 and constituting the Little Ice Age, was responsible for a wide range of strange occurrences: glaciers from the Swiss Alps crushed entire villages; the New York Harbor froze, allowing New Yorkers to walk from Staten Island to Manhattan; Eskimos inadvertently landed their kayaks in Scotland. One consequence was that trees grew much more slowly and produced tree rings that were smaller and densely packed. Some scientists—climatologists in particular—maintain that Stradivari took advantage of this wood in constructing his instruments.

Perhaps it's the way the wood was crafted? Surely the master fine-tuned his instruments by shaving the top and bottom to a variety of minutely differing thinnesses and thicknesses. He perfected the bridge, which the strings rest on as they cross the body. His bracing,

Dozens of completed and nearly completed violins and violas made by students and professors at Cremona's International School for Violin Making hang in a storage room.

bass bar, and soundposts were perfectly positioned. But Stradivarius violins have physically been reproduced exactly but without yielding the exact reproduction of sound. There are over a hundred violinmakers in the town of Cremona alone, and none has been able to unlock the secret. Scolari explains through our interpreter, "We can take all of the measurements of a Stradivarius *exactly*, and use machines to reproduce a violin *exactly*, but each piece of wood is different. The maker must feel the wood." And no one seems to feel the wood in the exact same way as the master.

Stradivari mentioned yet another possible variable in a letter to a client who was inquiring about the delayed delivery of his instrument. He wrote, "It requires the strong heat of the sun to bring the violin to perfection."[8]

The person who has perhaps come closest to imitating the perfection of a Stradivari violin is the aforementioned Joseph Nagyvary. His research knows no bounds. He's experimented with woods by soaking them in urine (at one point collecting the required chemicals by hanging a sign, reading "Please contribute generously to violin research," above a soaking tub in the biochemistry lab's men's room), crafted varnishes using ingredients ranging from ox bile to ground quartz, and used sound analyzers to discover the frequencies of Stradivari's soundboards. He states, regretfully, "What I wish I could have done is get more authentic samples of varnish for analysis, but I have not received any new samples in fifteen years. The dealers sent out word—anyone who owns a violin was warned not to let me close to the violin because I may even scratch off some varnish with my fingernail!"[9]

But in the end, the thing that makes a Stradivarius a Stradivarius is the same thing that makes a Monet a Monet. Its beauty is indefinable. It can't be analyzed in pieces but must be taken as a whole: age, wood, wonderment, craftsmanship, and all. Violinist Itzhak Perlman responded, when asked if he knew what was behind the secret of his Strad, "No, absolutely not. It's wonderful, you know, to have some sort of mystery that cannot be explained, cannot be put in a computer and analyzed. It's nice."[10]

THE MAKING OF SWEET BABY JAMES'S GUITAR

I've been in James Olson's shop for only three minutes and have already committed a near-grievous sin. The lanky fifty-something-year-old craftsman has taken down one of fourteen guitar bodies hanging from specially designed brackets outside his finishing room to show me some details. I reach out. My first instinct is to caress it; to run my hand along the curvaceous Brazilian rosewood body, to rub a finger across the mother-of-pearl inlay around the sound hole. And just as I'm within inches of the caress, Olson ever so politely says, "I prefer you not. The oils on your hands can mess up the finish. See." And sure enough, there's an oily thumbprint on the otherwise perfectly sanded spruce top. "A guy in here yesterday touched this one, and now I have to resand it. After sanding, it's white gloves only."[11]

And when I finally do get to fondle and strum one of his completed guitars, I understand the precautions. From the shave-in-the-surface eight-coat finish to the tone it emits, an Olson guitar is as close to perfection as a guitar can get. I'm not the only one who thinks that. James Taylor, David Crosby, Sting, Kathy Mattea, and Lou Reed are just a few who like to play Olsons.

At first blush it seems that the $12,500 base price for one of his guitars is excessive—until you spend an afternoon with James. You find out that a single "set" of Brazilian rosewood—the two pieces of wood required to make the back and the two pieces bent to make the sides—can run upwards of $1,000. He shows me a short stack of rosewood—a pile just barely larger than a good-size fireplace log—and tells me he's just paid $2,400 for it. "They're nice boards, but not perfect," he reveals.

In his storage loft and upstairs assembly room there are other piles of wood, just as gorgeous, rare, and valuable: a stack of ebony fretboards, stacks of German spruce for tops, ziricote, African mahogany, koa, and more rosewood from other countries for the sides and back. Most of the woods are quartersawn—cut along the log's radius—because woods cut in this direction are the most stable. Other cuts of wood may have better figure and prettier colors, but they're more prone to expansion, contraction, and cracking.

It may be a toss-up between which things Olson loves more: his wood stash or his Fadal, a CNC (computer-numerical-control) VMC (vertical machining center), which can, will, and does automatically make anything its owner can draw with CAD software. It's the size of a tollbooth, and inside it sits a router on steroids. In the back is a turntable holding twenty-one different bits, each already grasped in its own special chuck.

James points to a small screen displaying part of the computer program for crafting the necks, and begins explaining. "The computer drawing will tell it to grab tool number seven, run at a spindle speed of 10,000, geo or rapid move to X 11.9558 and that Y number. Then it's going to feed down to point 55 and cut out that inside, then it'll grab tool number 4 and clean up the corners that the bigger radius bit

James Olson, with one of the forty to sixty guitars he handcrafts each year, in front of pictures showing some of the musicians who play his $12,000+ instruments. James Taylor, Sting, Kathy Mattea, and David Crosby all own and play Olsons.

couldn't get, then it'll grab a 30/1000ths bit and clean up the corners." And that's just the start of the program, which holds 10,000 lines of code to shape just the back of the neck. He's already shown me the software program on his computer screen upstairs, which not only rotates the finished neck in three dimensions in an infinite number of positions but also gives a virtual tour of each bit as it cuts and routes its way through the virtual part.

It's not the romantic notion most people hold of a luthier. A luthier is supposed to sit at an old workbench, with piles of plane shavings surrounding his feet, jars of mysterious varnishes perched on the shelf, cat on the overstuffed chair, one unique-sounding instrument at a time being created by gut and instinct.

"All last week I spent twelve-hour days with a respirator strapped to my face, sanding. Before that I spent days pulling off little pieces of tape that hold the binding edges in place until my fingers were shot; some guitars have six hundred little pieces of tape. The end product is romantic, but not necessarily the process." There's a lot of handwork in the fifty to sixty guitars Olson crafts every year, but if it's possible to create a jig to make a part more accurately or a process faster, James has made it. He has jigs for creating the bracing, for cutting the kerfing strips that hold the top to the sides, for routing the dovetail where the neck slides into the body. Before buying the Fadal, he had seventy different routers, each with its own setting, its own bit, and its own highly specialized job.

He's not afraid to improvise. One router he still uses for working on arched guitar tops and backs is mounted to an old bicycle fork. His mammoth, multiton, stationary belt sander is concocted from salvaged conveyor belts, beer keg parts, and army surplus switches. It's more accurate than anything he could buy.

And soon you learn that technology used in conjunction with artistry doesn't diminish the quality of a guitar but enhances it. The two together create a better product than either method alone could create. If you've created a guitar with a perfect tone, the goal should be to use whatever tools and techniques are necessary to accurately re-create that instrument and tone over and over again. Part of artistry is creating good jigs, good computer programs, good quality control.

And part of artistry is staying in business, or at least surviving long enough to develop the art. When Olson started, there simply wasn't any good information available on guitar making. While trying to figure out how to build curved guitar sides he sought help from a vocational school instructor, who told him they were probably bandsawn out of solid chunks of wood. Olson rolls his eyes. "You know what two chunks of rosewood 4 by 8 inches would cost? If you could find it?"

But he kept at it, kept getting better, and kept trying to build a business. He worked as a janitor in a church, where he swapped mopping floors for workshop space. "I was cleaning toilets with rubber gloves on when James Taylor called." It was a call that eventually launched his business into the stratosphere.

At one point, Olson had backorders for 160 guitars—a 4-year-long waiting list. Through pricing and learning to mouth the words "I'm not taking new orders now" he's gotten his business and his life back under control.

When asked what makes a good guitar, he says, "A good guitar is always right on the verge of falling apart. They have 135 pounds of pressure pulling on them day and night, so they need to be braced well, but still be responsive. They need to be built solidly, but not so solidly that you lose the tone." And life for a guitar traveling on the professional level is tough. It goes from 80 percent humidity at an outdoor concert to 20 percent humidity in an air-conditioned dressing room. A humidifier gets thrown in for a few days, then forgotten about for a few days. The spruce top and the braces contract at different rates. "It's tough out there for a guitar," says Olson.

At lunch I tell him he seems to have the perfect life—all the work he can handle, a state-of-the-art shop 60 feet from his back door, rubbing elbows with famous musicians on a regular basis—the things people dream about. But he responds that artists—guitar makers included—are really insecure. "You can't stand the thought of being left behind or passed by. You want to stay on top, and then once you're on top, people start criticizing your guitars and try to bring you down. It's like being a shark. You can't stop, or you sink to the bottom."

Olson is miles from the bottom.

DRUMS: AND THE BEAT GOES ON AND ON AND ON . . .

When one is searching for the earliest wood musical instrument, one must first define the word "music." Good luck. What was music to me was "turn that crap down" to my parents. What's music to my kids is "turn that crap down" to me. American composer John Cage penned a piece titled "4'33"" in which the performer sits in front of the piano for four minutes and thirty-three seconds without playing a note. On the other end of the spectrum sits Lou Reed, who, on his 1975 double album, *Metal Machine Music,* delivers a solid hour of abysmal guitar feedback that prompted *Rolling Stone* music critic James Wolcott to liken it to "the tubular groaning of a galactic refrigerator" and "spending the night in a bus terminal." Most of us will find our own definition of music somewhere in between, but surely one element most people will include is rhythm, and that is where it all started.

The development of music predates the written word by tens of thousands of years, and few ancient instruments—especially of wood—have survived; there's really no telling how things progressed. Chances are that music began with readily found objects such as sticks, bones, or dried fruit with seeds that rattled.[12]

Studies of primitive cultures point to stamping sticks as the very earliest instrument; most were simply extensions of dancing feet or clapping hands. Solomon Islanders created mighty thuds by dancing over wood-covered pits, and the Balinese killed two birds with one stick by grinding corn as they rhythmically banged poles into a wood trough full of it.

Eventually, found objects were purposefully crafted into instruments to create specific sounds and, in many cases, melody. One can imagine the wind playing a monotone or two on a washed-up conch shell and humans elaborating on that. Animal horns became musical horns, hollow sticks had holes drilled in them to become flutes, and animal bladders became bagpipes.

In terms of the first "made" musical instrument, most indicators point to the drum, since all that was needed was a log. The earliest drums were hollowed-out logs, which were banged upon. Eventually, drums were crafted in a huge array of options. The shell or body of the drum could be made of wood, metal, pottery, tortoise shells, or, in at least a few cases, human skulls. They could be cylindrical, conical, hourglass, bowllike, and cigar-shaped. The heads or membranes could be animal skins, bladders, parchment, or cloth. The heads could be secured by cords, bands, buttons, nails, or glue. They were played with hands, sticks, mallets, wire brushes, rattles, and bones.

Excavations in Mesopotamia and Germany have unearthed small drums dating back to 3000 BC, and other evidence dates them back to at least 6000 BC.[13] Drums were handy things to have around. They were a wonderful way of communicating over long distances and excellent at keeping evil spirits at arm's length. Ancient Greek soldiers used to drum on shields made of stretched oxhide to frighten their adversaries. And drums were an excellent way of keeping galley slaves in synch as they rowed.

One of the oldest and longest-enduring drums is the taiko, a wood cylinder, traditionally carved from a single piece of wood, with cowhide stretched over either end. In feudal Japan, taiko drums were

Gigantic taiko drum made from a single piece of wood from a twelve-hundred-year-old tree. The drum weighs 3 tons, and its diameter is nearly 8 feet.

used to bless seasonal crops, summon rain, and drive away harmful pests. They were used to set the pace as soldiers marched into battle, as well as to issue orders on the battlefield. Now in America, taiko drums are used to entertain.

Contemporary taiko drumming is alive and well—and extremely impressive. One troupe totes a drum carved from the trunk of a twelve-hundred-year-old tree, with a length and diameter approaching 8 feet and a weight of 3 tons. Taiko drumming can also be vastly entertaining. Many performances end in a sort of "dueling banjos" competition on the big drum with performers on either end of the instrument aerobically endeavoring to outdrum one another.

Taiko drums today are crafted using both modern and ancient techniques. Many are made of staves rather than a single piece of wood. Drumhead stretching is serious business. Hair is removed from a fresh cowhide and placed over the drum shell. The drum shell rests on a circular platform with four to eight hydraulic jacks beneath it, depending on the size of the drum. The edges of the cowhide are secured to a framework below the jacks, and then the jacks are raised in small increments to tension the drumhead. The scene of a skin-stretching ceremony looks like a love fest held in a tire shop.

Drums are grouped under the heading of membranophone, and while most are hit, some work by friction. The most basic form of friction drum is a pot with a membrane stretched over it, through which a stick is moved up and down. The membrane could be anything from a pig bladder to an animal skin, and the "drumstick" anything from a stick to one's finger to a clump of horsehair. Archaic sounding, yes, but at a recent performance of *Stomp*, an entire song was played by two musicians sitting in a pile of trash and sliding plastic straws in and out of plastic-lidded Burger King paper cups—not something most people would want to listen to regularly, but nonetheless the recipient of thunderous applause.

And the beat goes on.

THE STEINWAY D: TWELVE THOUSAND PIECES OF INDESTRUCTIBLE MUSIC

Wu Han and Gilbert Kalish have just finished putting their piano through a torture test that's the musical equivalent of the 24 Hours of Le Mans. They've played Igor Stravinsky's "The Rite of Spring," a score which, when one of Stravinsky's contemporaries first heard him play it, remarked, "Before he got very far, I was convinced he was raving mad."[14] Rather than playing the "two pianos, four hands" version, Han and Kalish made the decision to rehearse and play the song using the "single piano, four hands" version in the spirit which Stravinsky originally conceived the piece. When Han announces this before the performance at Lincoln Center's Alice Tully Hall, the knowledgeable ones in the crowd ooh and aah as if she'd just pulled a rabbit out of the soundboard.

The score involves eggbeater trills and relentless hammering. Han plays the bass notes with such passion that the combination of her hands pounding the keys and her foot hitting the damper pedal launches her airborne time and again. Kalish, though twice her age, is equally impassioned. This is not tennis, where the veterans convert to doubles because they've lost a step; Kalish holds his own. At the end of the thirty-four-minute performance, the two stand, grasp hands, bow, droop off the stage, and then return to two standing ovations. They've left it all on the floor.

A conservative estimate shows their twenty fingers have dealt the eighty-eight keys a cumulative twenty-five thousand blows over the course of the piece. Each of those twenty-five thousand key strikes involves the Rube Goldberg–like chain reaction of fifty-eight different components that constitute the action for each key—EACH KEY! And through all of those million and a half intricate interactions, there is not one slip, stick, twang, or tick. It's a Steinway Model D concert grand: the grandest of the grand pianos.

* * *

The next morning, a dozen people, including my wife and me, tour the Steinway factory located on the edge of a working-class neighborhood in Queens, New York. Bob, our host for the morning, tells two stories that further exemplify the indestructible lineage of a Steinway.

He stands beside an army-green upright called the Victory Piano. It is one of three thousand pianos made by the factory during World War II (this in addition to the gliders it built for carrying troops behind enemy lines). "This piano had the unique ability to be air dropped," explains Bob. "So during the Battle of the Bulge, when morale was low, what did the government do to show they really cared about you? They parachuted in a piano. There are pictures of troops in full battle array singing around the piano. The shipping cases were so strong that some of the troops actually lived in them temporarily when they were on the front lines. The pianos came complete with tuning tools and sheet music."

An earlier factor prompted Steinway to create brawny pianos *prestissimo.* In the 1850s, when Henry Engelhard Steinway first began building pianos, most American-made instruments were poorly built. Four or five pianos would need to be placed on the stage for a concert because the instruments would literally fall apart as the musician played. "The piano would actually start warping and delaminating as they played. Strings would pop, and the musician would move from instrument to instrument right in front of the audience," explains Bob. "The mid-nineteenth-century pianists were the original music superstars. They would be up on stage for hours on end, and a piano would fall apart, and the women would swoon and throw flowers on the stage. And the audience loved the stuff."

But Henry Steinway didn't. He set about building a piano that would not only withstand the passionate pounding and 40,000 pounds of string tension to which pianos were subjected, but also project better in the larger concert halls of America. He started with the curvaceous outer rim, the largest and most visible part of the piano, the house in which every other component lives, constructed in the area of the factory where the real tour begins.

In an opportunity that brings me no small joy, the tour is timed so

Once the glue is applied to each of the eighteen wood strips, this small platoon of workers has twenty minutes to cajole, clamp, and convince this "book" of hard maple strips that the rest of their lives would best be spent as the rim of a Steinway grand piano.

we can witness the bending of an outer rim, or case. Not just any rim, but the rim of a Steinway concert grand, Model D. When Bob is asked about the model designations, he explains that models S, M, L, and B stand for small, medium, large, and big. And D stands for damn big. And with a length of 8 feet 11 3/4 inches, he's right.

On the lowest level of the four-story Steinway factory, the door to a freight elevator opens and six men file out. In the same manner they might carry a gigantic boa constrictor, each is toting a portion of a 20-foot-long stack of wood, known in the plant as a book. This book consists of nine 12-inch-wide and nine 6-inch-wide pieces of 3/16-inch-thick hard maple. The wider pieces will become the outer rim; the narrower pieces, the inner ledge that will support the soundboard. The men lay their cargo on a long skinny table and go to work. The process has a practiced, precision formality that oddly resembles a military funeral, without the sadness. There are no wasted motions. Few words are spoken. Everyone knows exactly what to do and when.

They run each strip through a machine with rollers that spreads an even layer of amber-colored glue on both sides (except for the two outermost strips, which get glue on just one side). As each piece rolls by you can see it is clear, straight-grained, and knotless. Because of the piano's size, most pieces are spliced together end to end to create the needed length. The men restack the gooey strips in the proper order as they emerge from the machine. Then, again in unison, they pick up the strips and march over to the rim-bending jig. It's 9:55 a.m. They have exactly twenty minutes to convert this 400-pound sandwich of wood and glue into the rim of a Model D, the instrument Han and Kalisch attempted—unsuccessfully—to beat to death the previous night.

The form they bend the strips around is, unsurprisingly, the shape of a grand piano, and the system and tools they use have remained basically the same since 1870. They begin by placing one end of the book vertically along the straight site of the piano form: the bass side. They position a strip of metal over the outside of the wood sandwich to help protect the outer veneer and more evenly distribute the pressure, in preparation for the clamping process. They position large blocks of wood, which mimic the outer shape of the rim, between the clamps and the strips, so that when the clamps are tightened the strips of wood have no choice other than to conform to the jig. The heat and moisture in the glue have relaxed the wood fibers to make them more pliable. The long straight side is clamped and tightened into the form. To gain the proper leverage in tightening the clamps, they use tools resembling 5-foot-long socket wrenches and bicycle handles. There are no gauges or guides to determine the correct clamping pressure; everything is done strictly by touch, look, and feel.

With the straight section clamped in place, the men who have been holding the free end of the eighteen strips make a running charge toward the form in order to prebend the strips so the clamping procedure can continue. The process of positioning the outer forms, then tightening the clamps, proceeds like a choreographed dance. When they get to the inner curve of the piano, one of the workers breaks out a 6-foot-long wood pry bar to coax the strips into the convex shape.

It's 10:10 by the time they clamp the final outer curve that completes the treble side of the piano. By 10:12, the final clamp is tightened. Seventeen minutes have elapsed.

Smaller piano rims remain in their form for six to eight hours; rims for the Model D sit overnight. The gestation period of a Steinway, like that of a human, is nine months. And like babies, no two are the same. The better part of the first trimester is spent in the rim-conditioning room, where the glue cures and the wood becomes accustomed to a shape it never could have conceived of as a tree.

If you subtract the weight of the 350-pound cast-iron harp to which the strings are attached, a Model D tips the scales at 640 pounds. Of that 640 pounds, 90 percent is wood, and 10 percent is everything else. Warren Albrecht is the person in charge of procuring that 90 percent, and for all the Steinways produced, it adds up to just shy of 1 million board feet a year.

Five basic woods—maple, yellow birch, sugar pine, poplar, and Sitka spruce—are used in every piano; more in custom-made or limited-edition models. When a piano costs $30,000 to $90,000, you use only the best

Albrecht sources the hard or sugar maple used for the rim from the New England area—the same area that's been supplying it for the past hundred and fifty years. But instead of floating down the Atlantic coast and into Long Island Sound in the form of huge log rafts, as was done in the nineteenth century, the wood today arrives by truck. It comes from the same maple tree tapped to supply your pancake syrup in the morning. An even harder maple, one called action maple in the plant, is used for some of the smaller components of the piano's keyboard action.

The yellow birch also comes from the New England area. Its tight grain structure, strength, and smoothness make it the perfect wood for the inner cores of the piano hammers. These same qualities make it a candidate for use as legs and pedal lyres.

Poplar is stable, straight grained, and smooth making it an excellent core for veneered components. It's used for the tops, keylids, and other flat surfaces.

Sugar pine, another stable wood, has a high strength-to-weight ratio and is used for the strips that brace the piano's soundboard.

But Albrecht spends the lion's share of his time seeking out and buying the $2 million worth of Sitka spruce used yearly for the soundboards. His beloved Sitka spruce grows along the west coast of North America, the bulk of it coming from Alaska and British Columbia.

Sitka spruce is the third tallest tree species in the world; the tallest known specimen stretches 315 feet, and the largest contains 10,540 cubic feet of wood. If Albrecht had access to this latter tree—and it generated nothing but pure, straight-grained boards—he would have enough soundboard material for five thousand pianos; he could twiddle his thumbs for two and a half years. But it doesn't work that way. Even after he's visited the Pacific Northwest mills and sorted through the wood, often literally board by board to find the finest, the people assembling the soundboards a year later may reject fully half of what he's purchased.[15] Any wavy grain, knots, or pitch pockets are grounds for ejecting a board from the game.

When asked whether it's becoming more difficult to find good soundboard material, Albrecht thinks, yes. "From the available sources there should be Sitka available, but the quality may not be as good. This is due to logging in areas where the quality won't be as high, like on the sides of hills. Most of the large prime logs have been harvested over the last many years, and now they're logging secondary logs of smaller diameter."[16]

It's the soundboard's job to vibrate and transmit every note ranging from the lowest A to the highest C and all eighty-six notes in between. It is the soundboard—each with its own minute variations in how the wood was grown, dried, cut, glued, shaped, and eventually installed—that helps create each piano's unique voice. The rosewood veneer on the side of a piano may mesmerize, the poplar for the top may be flat as a mirror, the walnut carving that graces the curves and legs may be downright Michelangeloic, but unless the soundboard—the piano's amplifier—is made of the purest, straightest-grained Sitka spruce, everything else is just window dressing.

When the tree is milled, the boards are quartersawn, the pieces fanning out from the center of the log like long rectangular slices cut from a pie. Quartersawing generates boards that are strong, straight-grained, and less prone to warping but, more important, resonant. With this type of milling it's rare for a given board to exceed 6 inches in width, meaning that ten or more boards are glued up to create each large panel. "The perfect board has anywhere from eight to thirty or more growth rings per inch," explains Albrecht, handing me a scrap from the floor, one that I later count to have thirty. "The closer the growth rings are to ninety degrees to the surface, the better. Lots of growth rings are important, but grain orientation is *really* important." It is this same high-quality Sitka spruce that is sought out by many makers of violins, guitars, harps, and another acoustical instruments. Albrecht is not without his competition.

Creating the half-inch-thick soundboard begins by dry-assembling boards to create a 5-foot by 5-foot panel. The grain is oriented at an angle to the keyboard, roughly mimicking the angle of the inner curve of the case. This translates into longer stretches of grain—nearly 2 feet longer than if the boards were oriented perpendicular to the keyboard—which in turn translates into better resonance. The assembled but still-loose boards are cut to the soundboard's rough shape. Glue is applied to the edges of the individual boards, which are then reassembled on a machine that's part clamping table, part paddlewheel.

Later, arched ribs are glued to the underside of the soundboard for strength. The bridge—which the strings will ride across to transfer vibrations to the soundboard—is installed next. Then the outer perimeter is thinned to further increase resonance. Each soundboard is a unique blend of man and machine, wood and glue.

Steinway is a company that has struck a balance in many respects. There is easygoing banter among the workers, but little standing around. The factory is loud, but not so loud that workers—most, anyway—wear hearing protection. Aging veterans, some having been with Steinway for thirty-five years, work alongside rookies. When machinelike consistency creates a part—like the hammer shank—with more accuracy, it's made by a CNC router; when that hammer shank needs to be tweaked into perfect alignment, it's heated by flame and twisted ever so gently by hand. The factory is the twenty-first-century embodiment of the ideals eschewed by the nineteenth- and twentieth-century Arts and Crafts movement: man and machine should be partners. Craftsmen use machines to do the mundane, which in turn frees them to apply their skills to things that can best be done with the human touch.

At Steinway, even setting up an interview requires that human touch. My initial flurry of e-mails garnered no response, but when an acquaintance informed me that Steinway still did things the old-fashioned way, I called company headquarters, and a real person answered the phone—on the second ring. When I asked about setting up an interview and tour, the receptionist said "Here, talk to Loretta." We chitted and we chatted, and by the time I'd gotten off the phone, I felt that if I were lost in Queens on some cold winter night, I could call Loretta at home and she'd invite me over for a cup of steamy hot chocolate.

The construction of a Steinway follows two somewhat independent tracks: the building of the case and the building of the key action assembly. Part way through the process the two tracks merge, and the task becomes one of creating a successful marriage.

The case-building track begins with the rim, then moves on to fraizing, whereby the top and bottom of the case are trimmed evenly, and the ends evened up and trimmed to their final dimensions. The maple wrestplank, which will hold the string pins, is doweled between the free ends, and the basic shape of the piano is established. Next, a pattern of hefty internal bracing made of Sitka spruce is installed. Here the tonal quality of Sitka spruce isn't important; but its high strength-to-weight ratio is.

The top, which is an enlarged version of the soundboard, is crafted from poplar, then veneered according to the final finish: maple for pianos that are ebonized, other veneers for those of natural wood. The hinged fallboard, which covers the keyboard, the music stand, and the desk, are made of similar materials. Legs for ebonized pianos are made of birch; those for most natural wood pianos—regardless of species—are normally walnut.

Limited-edition and custom-crafted cases receiving hand-carved elements are wheeled to a corner of the factory, where—oddly enough—they are hand carved. Wood is soaked, bent, applied, then hand carved. Legs that have been roughed out on a lathe are fluted and claw-footed. It is Old World craftsmanship right down to the wood shavings on the floor and the German accent in the air.

The keyboard action track begins in another part of the factory. There are fifty-eight parts to each key: some wood, some felt, some metal. Wood parts include the whippen, the hammer core, the hammer shank, and smaller connecting pieces; each needs to fit with the tolerances of a Ferrari valve job. Looking at a small model of a key, I find nothing intuitive about the way it works. But it does. When I ask Albrecht how fast the action on a key is, he tells me, "No human being has ever been able to play faster than the action can return it." Eventually, all five-thousand-plus keyboard components are screwed, glued, stapled, fitted, and twisted into a single keyboard-and-action assembly.

In another part of the factory, the various elements methodically begin to converge. There are twelve thousand parts, which, in Bob's assessment, makes it "the most complex, handcrafted device in the world." The soundboard moves to the belly department, so named because this is where the "guts" of the piano are installed. The soundboard is trimmed to finished size based upon the exact case into which it will be installed. It is exactly fitted to the case surrounding it as well as to the metal harp that will sit over it. Through trial assembly and disassembly the bridge is hand chiseled and hand planed. Soundboard, bridge, strings, and harp are all fine tuned to create the perfect storm. Strings are strung, the top is hinged, the keyboard assembly is slid into place, and the system of tuning and fine-tuning and fine-fine-tuning continues.

Steinways are first and foremost instruments of musical perfection, but our assistant tour guide reminds us of something else. "When you purchase a Steinway you're purchasing not only an instrument but a fine piece of furniture as well."

Veneer is applied to all of the rims, tops, and other exposed surfaces regardless of the final finish. The majority—which are destined to be lacquered black—will receive hard maple veneer, which accepts the glassy finish well. But Steinway makes custom and limited-edition pianos in natural woods, and the storage room we tour is full of them. Piles of veneers are stacked in flitches, thin sheets of wood restacked in the order they were cut from the log. Our guide shouts over the roar of a nearby saw, "There's over $2 million worth of veneer in this room alone." Albrecht rolls his eyes. He's the one who has purchased every slice of veneer in the room. "That number's a bit high," he whispers under his breath. But not by much.

The most exotic veneers go into Steinway's Crown Jewel Collection, a line in which buyers have their pick of eleven different veneers: Kewazinga bubinga from West Africa, fiddleback anigre from Africa, East Indian rosewood from Sri Lanka, Santos rosewood from Belize, figured sapele from Nigeria, African cherry, bee's-wing mottled satinwood, Macassar ebony, African pomemele, and the more familiar and pronounceable woods like mahogany, cherry, and walnut.

Steinway also continues to make one-of-a-kind and few-of-a-kind art case pianos, as the company has since 1857.[17] Some of them are so over the top that they make Liberace's mirrored grand piano look like a Yugo. The St. Croix, clad in copper-red Cuban mahogany with satinwood accents, all but glows. The Pear Grove is a riot of marquetry with thousands of inlaid pieces of ribbon-striped bubinga, walnut, mahogany, and cherry. The Europa is clad in book-matched walnut and Carpathian elm burlwoods. Renowned woodworker Wendell Castle designed a mammoth art case piano in 1987 to commemorate Steinway's 135th anniversary.

Perhaps none of these will reach the fame and auction price attained by the Alma Tadema art case piano built in the late 1800s that recently sold for $1.2 million—but you can buy a recently built re-creation for $675,000.

Steinway is also fond of commemorative or limited-edition pianos. "It seems like every few years they come up with a reason to issue one," explains Bob. There's the Roger Williams Limited Edition Gold

Piano. Only eight were made; the music desk is decorated with notes from "Autumn Leaves," the best-selling piano recording of all time. Steinway issued the Tricentennial to celebrate the 300th anniversary of the invention of the piano and not one, but two, pianos marking Steinway's 150th anniversary.

On the second floor we encounter an example of the limited-edition piano they'll be offering next: the Henry Z. Steinway piano, celebrating the ninety-first birthday of the great-grandson of Henry E. Steinway; ninety-one will be made. The rim is finished in East Indian rosewood. "The veneer alone costs $12,000," boasts one of our guides. Albrecht rolls his eyes again and quickly does the math. At $8 a square foot and 350 square feet of veneer, the number is a lot closer to $3,000. But that doesn't make the piano any less stunning. The legs are massive barrel-fluted affairs, the decorative trim and medallions are hand carved, the music stand is intricately carved to include the initials HZS, and Henry Z. will personally autograph each cast-iron harp.

Today this American company makes about two thousand grand pianos and six hundred uprights a year. The first American-made Steinway bore a number in the 400s; each since has carried a sequential number. Today the pianos rolling through the factory bear numbers in the 577,000 range. "It's taken Steinway 153 years to get to that number," explains Bob. "That's about two years of production for some piano manufacturers today." The numbers remain useful. If you own a Steinway, regardless of age, you can call the company and give them the model and serial numbers, and they'll tell you who originally bought the piano, when it was shipped, and, in some cases, its repair history.

The factory in Hamburg, Germany, established by Heinrich Englehard's son, C. F. Theodore, in 1884, manufactures about a thousand pianos a year. The history is equally intriguing. The Hamburg plant was leveled by the Brits in 1944, but not before the workers squirreled away the bulk of the valuable veneers and woods for a more peaceful day. When the plant reopened in 1948, the loyal old work force showed up en masse, raw materials in hand, and began making grand pianos almost immediately. It is this type of Old World loyalty in which the Steinway factories on both sides of the Atlantic have been marinated.

In the end, you have a piano that is one of a kind—but one of a kind within strict parameters. The number of growth rings in the soundboard, the adjustment of the hammers, the subtleties of the case shape, the handcrafting done by different workers on different days give each Steinway its own unique sound.

A story told by an acquaintance who is also a concert pianist proves a case in point. Visiting the showroom of the Steinway Hall sales office, where upward of fifty grand pianos are kept, he sat down and hammered out a few measures. As he left, the salesman—who had been seated down the hall and five rooms away—complimented him. He knew not only which part of which song he'd played, but also exactly which of the fifty Steinways he'd played. Each piano has its own personality, its own DNA. Perhaps Bob is right in his assessment that "the only competition for a Steinway is a Steinway. Every other piano is merely a piano-shaped object."

THE NATIONAL MUSIC MUSEUM: SIX HUNDRED ZITHERS, B. B. KING, AND ONE-TON DRUMS

If I were to blindfold you and lead you into room No. 1 where four Stradivari instruments stand—including one of only two guitars crafted by the master, and the Harrison violin—you might think you were in Florence, Italy. If instead I were to lead you blindfolded into room No. 2, where a 1952 Les Paul and B. B. King's Lucille guitars sit, you might think you were at the L.A. Hard Rock Café. And if I were to take you to room No. 3, where two upright basses stand—one made from a cracker barrel, the other from a perforated duct pipe—you might guess you were in a Deep South backwater bar. And finally, if I were to bring you into the lobby, remove your blindfold, and show you the one-ton drum mounted on wheels, you might think you were at a Ripley's Believe It or Not Museum.[18]

But in all four cases your guess would be wrong. Because all four displays are in a single building, and that single building sits on a quiet street in Vermillion, South Dakota, on the campus of the University of South Dakota. The name of this unique place is the National Music Museum, and it is this diverse orchestra of instruments that John Koster, curator of the museum, is charged with overseeing.

Koster is a man with a very large endowment and a very small facility. The entire collection of thirteen thousand instruments—one of the largest in the world—is valued in the tens of millions of dollars; yet the museum occupies nine modest-size rooms of an ex-Carnegie library—a space that allows it to display only 10 percent of its collection at any given time.

Despite the storage crunch, instruments keep rolling in. The museum's quarterly newsletter lists the latest acquisitions; a recent list includes such diverse instruments as a 1937 Hohner harmonica played by "Karl A. Johnson, a dairy farmer and carpenter," a 1730 viola da gamba by Antonio Stradivari, and a twentieth-century Yugoslavian one-string folk fiddle called a gusle.

At one point during my visit, we walk through a back room with table after table of bubble-wrapped "somethings" carefully stacked. When I question Koster about the contents, he simply replies "zithers." Five to six hundred zithers, to be exact. Two collectors have meticulously wrapped their lifetime collection of zithers—some utterly fabulous, including a Franz Schwarzer zither inlaid with gold, ivory, and abalone—and bequeathed them to the National Music Museum. Koster will eventually find the space to store and display some of them. He'll somehow find the funds and personnel to catalog them, but it illustrates the challenge.

A curator's work is never done.

Koster's office consists of a desk tucked in the back of a storage room surrounded by piles of boxes and tarped-over instruments. He can find only a collapsible camping chair for me to sit on, and this leads me to believe my visit is somewhat of an exception. I look for a place to rest my coffee and begin to set it down on something with black plastic draped over it, but I halt, not knowing what's beneath.

"Maybe you should set it down on this box instead," Koster says, perhaps not as frantically as he should. "That harpsichord is 399 years old." The instrument, I later find out, was built by Andreas Ruckers, considered by most to be the premier harpsichord builder of all time. This is a good idea indeed.

Prior to meeting Koster, I had envisioned someone with a bowtie and accompanying manner; instead, I meet Beethoven in blue jeans. He has long graying hair and a face that exudes enthusiasm. Several times as we walk through the museum, Koster steps up to a vintage harpsichord and, assuming a stance similar to one Jerry Lee Lewis might use when playing "Great Balls of Fire," bangs out a few measures of some baroque song. He is a man enamored with music and absolutely smitten by the harpsichord. And though he's ever so down-to-earth and unpretentious, it soon becomes clear that Koster might as well have the word "Mensa" stapled to his forehead.

I ask him how, in a single building where there are wood instruments ranging from the $3 million Harrison violin to a 50-cent cigar box banjo, he knows what temperature and humidity to maintain the building. I expect an exacting answer—something with two numbers to the right of the decimal point—but instead get a "50 percent plus or minus whatever." And soon I begin to understand that neither Koster as a curator, nor the National Music Museum as an institution, views musical instruments as sacred shrines but rather as things to study, learn from, and revel in. The atmosphere is no stuffier than that of the Looking Glass tanning and wedding shop across the street. And while you can't pick up the Jakob Stainer violin and play "The Devil Came Down from Texas," you can dwell and linger and photograph (with flash) to your heart's content. You can push your nose to within an inch of the instruments on open display, and no alarms go off. There are hands-on displays that show how different key mechanisms work.

Weekly concerts involve such varied performances as "Front Porch Stories and Blues" by Alvin "Little Pink" Anderson and "There's No Business Like Bow Business" by the Dakota String Quartet. The eighteenth-century harpsichords in the concert room are played regu-

John Koster, curator of the National Music Museum, shows off an African drum that recently arrived, unpackaged, with the FedEx label stuck directly to it. "Sometimes," Koster explains, "restoration involves just getting the labels off."

larly, and during yearly conventions a select number of the instruments may be un-displayed and actually played. A few years back, the museum hosted a concert in which Eugene Fodor played the Harrison and three other rare violins. The staff could only hold their breaths as this musician—who's known for playing on the edge—stood on the edge of the stage bowing and gesturing with two priceless instruments chucked in either hand.

When asked to weigh in on the debate over whether great instruments need to be played frequently in order to retain their richness in sound, Koster is skeptical. "It's mostly romance and imagination," he explains. "In the past there was some truth to that [idea] because if an instrument was lying around for years, wood worm larvae could go about their business undisturbed. But if you took it out and played it once a day, [the worms] are not going to find it very hospitable and would leave. But presumably we're pest free."—a presumption Koster discovers to be wrong by the time I leave. And when I ask him about how hundreds of years of aging purportedly improve the rich tone of wood instruments, he remains a realist. "Wood oxidizes, and the long chemical bonds in the cellulose break down, and the wood becomes more flexible, which affects the tone," he says. "But most of that happens within the first fifty years, and with some instruments, like harpsichords, many of the changes that affect sound occur in a matter of weeks or months."

If one visits the museum focused on wood, it becomes clear that wood has been the raw material of choice for musical instrument makers throughout history. If you were to place all of the instruments in the Italian Stringed Instrument room on a scale, 98 percent of the weight would be wood. That number would be just as high in the Great American Guitars room and the Beede Gallery of non-Western instruments. This percentage would drop in the galleries containing the limited-edition Bill Clinton saxophone and the heart-shaped horn featured in "Sergeant Pepper's Lonely Hearts Club Band." But all in all, wood permeates the museum.

You discover ordinary instruments crafted of extraordinary woods: flutes crafted of solid ebony, clarinets of boxwood and crocus, pianos with purpleheart veneers. One harpsichord contains over twenty different types of wood, both functional and decorative.

And you see extraordinary instruments built from ordinary woods: The crwth (spelled correctly), a bowed lyre; the neverlur, a 6-foot-long horn covered with birch bark, used by herdsmen to frighten wild animals; an electric lapsteel guitar built in the shape of a crutch for Barbara Mandrell; and the serpent, a curvaceous wood horn that twists and turns like a Pike's Peak road.

I examine some 2-foot-long boxwood clarinets, which, when I compare the woods from which they were derived to the shrubby little boxwoods that have been struggling for years in our garden at home, seem impossibly large. But Koster explains that in days gone by, trees were cultivated over decades specifically to yield wood suitable for instruments. In areas of Switzerland, spruce are tended for *centuries* to produce tight, straight-grained wood for soundboards. But not all musical woods have been nurtured with such care.

Research by the SoundWood program, part of the Global Trees Campaign, indicates that as many as seventy of the two hundred "tone-wood trees" used in musical instrument making are of "conservation concern." Those most in jeopardy include Brazilian, Indian, and Honduras rosewoods used for stringed instruments and xylophones, African blackwood used for clarinets and other woodwinds, pau Brasil used for bows, and ebony used for fret boards. Preservation and sustainability efforts have a rough road ahead; few of the woods listed are tended as plantation timbers. And it's a long-range effort; some tone-wood trees require a hundred years to produce the mature wood coveted by instrument makers covet. But instrument manufacturers, including Gibson, Fender, Martin, and Taylor Guitars, along with other smaller companies, are on board with using certified woods. It's a start.

It comes as no surprise that Koster has a deep and abiding understanding of wood. He's taken a weeklong seminar from Bruce Hoadley—the high priest of wood, with a PhD in wood technology from Yale—in wood identification. And he uses what he's learned. While cataloging instruments for his book *Keyboard Musical Instruments in the Museum of Fine Arts, Boston,* he conducted over twelve hundred wood identification tests on fifty instruments and in the process discovered something on the order of fifty different woods.

Knowing how to identify wood is an essential skill for the true preserver, conservator, and restorer of wood instruments. If one is to do authentic restorations, one must know the starting point—and that starting point is knowing what woods the instruments are made from. Sometimes the accepted starting point is simply made up. In a catalog covering the Hill instrument collection at Oxford University, one instrument is described as having a cedar soundboard, and the description even goes to great lengths to explain that the tree grew on the shores of romantic Lake Como. "And it's all just garbage," explains Koster. "The soundboard is spruce." Some boards are wide enough to submit to dendrochronological dating—a comparative study of growth rings, which can help determine not only how old an instrument is but also where the wood came from, since ring growth rate varies according to climate and location.

Koster untarps the 399-year-old yet-to-be-restored harpsichord I had considered using as a coffee table and uses it as an example of wood varieties. The oak bracing came from the Baltic region, the soundboard came from Switzerland, the bridge is made from cherry, parts of the case were made from locally sourced poplar, and the keys were made from bog oak—oak that's been immersed in bogs for hundreds of years and turned black. "And each wood was originally selected because it performed a particular task particularly well," explains Koster.

There is another good reason for a conservator to know his or her woods: forgeries. While great strides have been made in the science used to detect forgeries, that same science can be applied in creating them. Ten years ago, several magnificently preserved seventeenth-century Spanish harpsichords surfaced, their previous whereabouts and pristine condition attributed to their having been sequestered in convents and monasteries for hundreds of years. After they were originally displayed as a group in a museum in Hamburg, Germany, the National Music Museum had the opportunity to acquire one. "It was shipped over, and when I first saw the instrument in the staging area, from a distance I noted the soundboard looked funny and was perhaps a reproduction," explains Koster. After donning his detective hat, he began to sense that the entire instrument, inscribed with the date 1641, might be fake. While the overall look of the instrument was quite authentic, the microscope told a different story.

The soundboard had been crafted out of boards each exactly 4 inches wide, which were discovered to be western red cedar clapboard siding from California—a product surely not available in seventeenth-century Spain. Thirty-six of the 180 jacks, the small strips of wood used to pluck the strings, were of a lighter color—an attempt to make the instrument look authentic by way of mimicking a common restoration done to truly old harpsichords. But close examination of the jacks revealed two things: ripples left by a modern rotary planer and slot shapes—identical in both old and new jacks—that had clearly been cut with an electric bandsaw. Magnification of the bottom surface of the soundboard liner revealed it had been cut with a circular saw. The painted decorations contained modern pigments such as titanium oxide. And perhaps the clumsiest part of the restoration was revealed when Koster discovered that the numbers used to mark the jacks had been made with a ballpoint pen.

There are various degrees of forgery. Many were perpetrated by twentieth-century Florentine music dealers. Some forgeries start with an authentically old instrument that is revamped to make it appear even older or to have been crafted by one of the masters. Others are conglomerates of several instruments, or crafted from pieces of wood from old cabinets in order to defy true dendrochronological dating.

One form of instrument alteration is accepted simply because it is part of the history of music in and of itself. Those alterations involved remodeling instruments in order to keep pace with the musical tastes of the day. We've already learned that only one of Stradivari's six hundred and fifty instruments existing today has been left in unaltered condition (indeed, only six maintain their original necks). Equally remarkable is the fact that if you were to survey *all* the violins, violas, cellos, violino piccolos, and other stringed instruments crafted by the Cremonese instrument makers of the seventeenth century—Stradivari, Amati, Guarneri, and others—you would find only three unaltered instruments on the entire planet.

The National Music Museum holds some remarkable examples of these alterations. The King violincello by Amati—one of the most delicious instruments you will ever see, with gilded and painted back—was reduced in size by removing a strip of wood from the center of the back. It was done in a manner that left the sword-wielding woman representing justice with no left arm or waist. It was also converted from a three-stringed instrument to a four.

We look at other impeccably altered instruments. We scrutinize one instrument in which the f-hole cutouts were moved to accommodate moving the bridge. You can squint and stare and find no trace of the splicing and recutting involved; the grain, the finish, the surface of the old and new match *exactly*. A bass viola da gamba made by Stradivari in 1730 was enlarged into a violincello a hundred years later. Again, the eye-shaped pieces spliced to the top of the instrument are not seamless, but beautifully done. Alterations aren't limited to the violin family. Koster shows me a harpsichord that he had personally restored back to its original harpsichord status after it had been converted to a piano many years before. He mimicked the original design right down to the segment of crow feather traditionally used to pluck the string.

As I prepare to leave, Koster shows me an African drum lying on a table behind his desk. It's indicative of the multitudinous ways in which instruments arrive at the museum. It's been shipped to him unwrapped and unprotected, with the FedEx and address labels stuck directly to the drum. It pinpoints the varied nature of conservation. "Sometimes," Koster jokes, "restoration involves just getting the labels off." He then looks a little closer at the drum and gasps, "This is not good." He reaches inside the drum and comes out rubbing his fingers together. He's spotted a dab of powder inside the drum:

powdered wood that wasn't there yesterday. Koster has discovered a wood-boring insect that has infiltrated the greatest music museum in the world, inside a nondescript African drum. As soon as I leave, he explains, he'll need to bag it and throw it in a freezer for a few days to kill the invader.

A curator's work is never done.

Stradivari violins, Steinway pianos, and Olson guitars provide the sounds that are music to the ears of concertgoers around the world. But different sounds are a different kind of music to other ears. For some, the crack of the bat, the smack of the pool cue, and the solid whack of the driver are the sounds that bring a standing ovation. Wood has found a solid place in the world of sports.

Here's a swing at a few of those stories.

CHAPTER 5

Wood in the World of Sports

Jim Anderson and Paul Johnson are giddily flipping through the pages of the latest *Sports Illustrated*. Not the swimsuit issue. They find the page they're looking for and hold it up to show off a full-page picture of Jimmy Rollins, shortstop for the Philadelphia Phillies and holder of a thirty-six-game hitting streak, the longest in the majors. The streak excites them. But what excites them more is the stick of wood Rollins has casually resting on his camel-hump right shoulder. Those *SI* readers who are willing to hold the picture up to strong light, and extrapolate the last four letters of the name on the bat, can determine the manufacturer. For those unwilling to do so, Jim and Paul will gladly tell you it's a MaxBat—and it came off the lathe of their small manufacturing facility in Brooten, Minnesota (population 649), just three months ago.

BASEBALL BATS: A DAVID-AND-GOLIATH AFFAIR

"When you're a bat manufacturer, there's no better advertising than when one of your guys goes on a hitting spree," explains Paul, the plant foreman, brushing off a morning's worth of sawdust. Unless it's your guy on a hitting spree, pictured in *SI* with one of your bats on his shoulder.

What makes this small bit of notoriety loom even larger is that the world of major league bat making is a David-and-Goliath affair. Goliath is Hillerich & Bradsby, maker of the Louisville Slugger—the official bat of major league baseball—with 70 percent of the pro market.[1] This year they'll make one million wood bats: a far cry from the seven million they made in the 1970s before the advent of the aluminum bat, but a further cry from MaxBat with an output of thirty thousand per year. But while Louisville has been around for almost a hundred and twenty-five years, MaxBats has been around for two. "Our goal is to be one of the top four manufacturers by the end of this season," exudes Jim, a former amateur baseball player and founder of the company.

But manufacturing capacity isn't the biggest thing differentiating the companies. While Louisville and most other manufacturers use the traditional northern white ash, MaxBats uses maple. Their slogan is "Our maple kicks ash," and Paul is more than willing to explain why. Maple has a tighter grain than ash, it's 5 percent denser, and it has greater surface hardness, which gives the ball more "pop" off the bat. The fact that Barry Bonds used a maple bat to hit seventy-three home runs in 2001 hasn't hurt business a lick.

Holding a disintegrating ash bat sent by a major league player MaxBats is using as a prototype for creating a maple version, Paul explains that while ash softens with use, maple becomes more compact and harder with use. Not just any maple will do. Maple that grows in Vermont and other parts of the northeast is too dense because of the long winters, and maple grown in the far south is too soft because of the long summers. But Pennsylvania maple—with just the right balance of density and resiliency—hits the sweet spot. The wood is cut in the winter when sap content is low, and only the most straight-grained sapwood is used.

The ash that Hillerich & Bradsby uses comes from a similar geographic location. Theirs is a 200-mile strip along the border between New York and Pennsylvania, "a wooded, rolling, mostly remote area blessed with just enough sun, just enough rain, and just enough glacial till."[2] Though large, the company continues to be picky when it comes to picking wood. They own 8,000 acres of timber, but only a small portion of the 10 to 12 million board feet of ash the company mills annually comes from those holdings. Of that vast quantity of wood, only 2 percent will become baseball bats. The perfect bat blank, or billet, will come from a 60-year-old tree, preferably one growing on a north-facing or east-facing slope, where the soil is richer, the trees more numerous, and as a result straighter and taller in their competitive bids to reach sunlight. That perfect billet will have straight grain and eight growth rings per inch, and it will come from the lower 10 feet of the tree. A completed 32-ounce bat starts with a 20-pound 40-inch-long piece of green ash. Eighteen pounds of sawdust and moisture later, a Derek Jeter bat will emerge.

History has clearly shown there's nothing wrong with bats made of ash. It is the sentimental favorite. In *A Natural History of North American Trees*, Donald Peattie writes, "Every American boy knows a great deal about White Ash wood. He knows the color of its yellowish white sapwood and the pale brown grain of the annual growth layers in it. He knows the weight of White Ash not in terms of pounds per cubic foot but by the more immediate and unforgettable sensation of having lifted and swung a piece of it, of standard size. He even knows its precise resonance and pitch, the ringing tock of it when struck." And of course this thing that "tocks" is an ash bat.

Ash is a team player. Nothing flashy in the grain structure, but straight and true is what you need when it comes to hard work. Ash trees are solid and dependable; you wouldn't hesitate having one watch your back. If they were human they'd vote conservative Republican, listen to Frank Sinatra, and bowl on Friday nights. But maple continues to eat away at the pie charts.

The earliest bats more closely resembled rolling pins, bottles, and other things you might find in the kitchen than the sleek engineered clubs of today. Until bat diameter was limited to 2 1/2 inches in 1859, you could find square bats, wagon tongues, axe handles, and sapling trunks being swung. Though they were well beyond that era, legend has it that Joe, Dom, and Vince DiMaggio made baseball bats from oars out of their father's fishing boat.

In 1869, a bat length limit of 42 inches was instituted, and in 1893 the allowable diameter was upped to 2 3/4 inches. But no weight restrictions were—or have ever have been—instituted. The average bat today weighs about 2 pounds, but before bat speed supplanted bat weight as the driving force, some truly monstrous bats were used in shows of remarkable brute force. "Home Run" Baker, who led the American League in home runs for four straight years beginning in 1910, used a bat weighing 52 ounces with a handle almost the same size as the barrel. At one point in his career, Babe Ruth used a hickory bat weighing 47 ounces, and at another point he ordered a 54-ounce bat, though there's no proof he ever used it on game day. Hall of Famer Ed Roush, the "Hercules of Heft," used short bats that weighed up to 50 ounces, sparking one historian to describe them as having "the appearance of well-sanded logs."[3]

They don't make 3 1/2-pounders at MaxBats. While they make several lines of standard bats for general consumption, they're focused on building a client base in the big leagues. Each player has exact specifications in terms of weight, length, and dimensions. Very exact. Dimensions are measured in fractions of grams and millimeters for the hundred or so major league players they supply.

They use a $100,000 computer-programmed lathe that can rough out a bat in ninety seconds. They program the computer with an exact set of specifications designed for the bat of each player. "The idea of a hand-turned bat seems real romantic," explains Paul. "But if you want absolute consistency, you need a computer." And you believe him when he breaks out a micrometer to check the dimension of the handle as a bat comes off the lathe. Paul and his production assistant, Matt Cooley, shoot for an accuracy of within 1/10 ounce and diameters within 3/1,000 inch of the players' specifications. They even take into account the thickness of the catalyzed lacquer finish—approximately 0.1 millimeter—in the final calculations. He hands me a bat blank fresh off the lathe, smooth as pumpkin skin—and that's before it heads to the sanding machine. Once off the sanding machine, the bat is brought to its final exact weight by slightly hollowing out the end on yet another special machine.

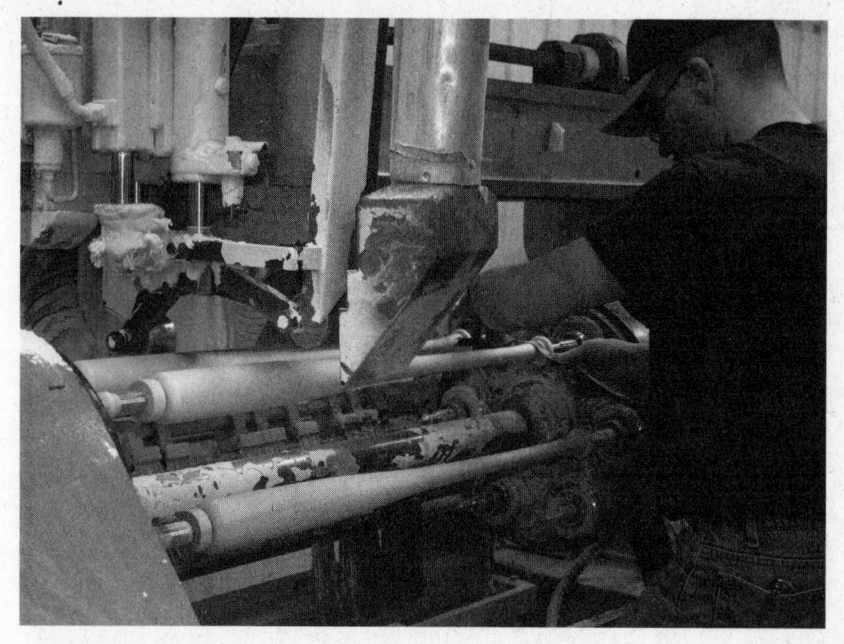

Maple bats being sanded after rough shaping on a computer-guided lathe. MaxBats—with the motto "Our maple kicks ash"—makes 970,000 fewer bats per year than Louisville Slugger but is slowly gaining market share.

This entire process is a far cry from how the first manufactured bat was crafted. As legend goes, Pete "The Gladiator" Browning, an outfielder for the Louisville Eclipses, broke his favorite bat during a game. A teenager in the stands by the name of Bud Hillerich invited Browning to his father's woodshop, where he took a chunk of white ash and began turning it on the lathe. Every minute or so, he'd remove the bat, hand it to Browning for a practice swing or two, then fine-tune it based on the slugger's comments. The next day, Browning used the bat and went three for three.[4] And though J. Fred Hillerich admonished his son, "There's no future in supplying an article for a mere game," the hundred million bats the company subsequently manufactured might prove otherwise.

Today the typical bat is 34 inches long, with a standard length-to-weight ratio of 1 ounce per inch. (If a player wants a 34-inch bat that weights 32 ounces, it's designated as a "minus 2" in the business.) Prior to turning, the MaxBat wood is dried in special kilns where radio frequency waves excite the water molecules and a vacuum process removes the moisture. This creates a wood blank in less time, with less checking, and with a more uniform look (yes, looks *do* count). Before the advent of the vacuum kiln, it was difficult to dry maple to a point where it was light enough to create a bat with the right length-to-weight ratio.

"These guys are serious about their bats," says Dick Johnson, president of MaxBats, who's joined us. "They can rub the handle and tell if the diameter is off 1/16 inch, swing the bat once to see if the bat's half an ounce too heavy or too light, ping the barrel with their finger to see if the wood's the right density. And if everything isn't perfect, it becomes a batting practice bat." And a strike against the manufacturer. Ted Williams had the same demanding specifics. He once returned a set of bats to the manufacturer with a note: "Grip doesn't feel just right." Subsequent measurements showed the bats to be 5/1000 inch shy of Williams's specified diameter. Most pros today order bats by the dozen and find three or four out of each batch that feel right and become game bats. The goal at MaxBats is for players to pick 12 out of 12.

That feel of the bat has always been an obsession for ballplayers, a traditionally superstitious lot to begin with. Eddie Collins used to season his bats in a pile of manure, Honus Wagner boiled his in creosote, and Joe DiMaggio rubbed his with olive oil. Babe Ruth liked bats with knots in them. Hall of Famer Cap Anson owned five hundred bats, many of them created from his penchants for spotting a particular type of fencepost, buying it from the farmer, and then having it turned into a bat.

Through the years, a variety of bats have come and gone. In 1884, a paper bat was developed, with predictable results. A few years later, the banana bat, shaped somewhat like a large spoon, was developed. The inventor rightly surmised that balls hit with the bat would spin, hook, and fly off the bat, making it "more difficult to catch the ball, or, if caught, to hold it."[5] Bedpost, knobby, and prebent bats were also developed, but in the end, the bat has remained a bat.

The obsession for the perfect bat has a basis beyond superstition.

Pinpoint accuracy—literally—is the name of the game. The collision of a bat and baseball lasts a mere 1/1000 second. If that fraction of a second occurs 1/100 second too late, the result, for a right-hander, is a foul ball in the right field seats. If the contact point on the bat results in a trajectory of anything other than a 35-degree angle, the ball will not travel the maximum possible distance. And if a bat has even a slight warp, it can be—and is—the difference between a home run and a routine pop fly.

Perhaps the earliest lesson in wood technology any of us received was taught via the baseball bat. When a bat contacts a ball on the edge grain (as opposed to face grain), a more solid connection is made. The face grain tends to compress, while the edge grain is much stiffer. Thus, the manufacturer's logo was always placed on the face grain, which—according to every Little League coach in the history of the world—should always face straight up or straight down when the batter is swinging. Failure to do so could also result in a broken bat, a scolding, or both.

Wood being wood, a bat's life span can vary drastically. Joe Sewell, a Hall of Famer, used only one bat through his entire fourteen-year career, chalking up 2,226 hits. But the thinner handles of today don't fare as well. Jim Rice once took a mighty swat, and his bat broke in half—the strange part being that he'd completely missed the ball. Paul at MaxBats swears that just as many bats get broken in the dugout by frustrated players as in the batter's box by hard-swinging hitters. "Bo Jackson used to break them over his knee." And on a good day, when a pitcher is jamming batters to the inside, he can send five or six bats to the wood pile.

Though injuries are rare, major league baseball requires that all bat manufacturers carry a $7 million insurance policy in the event a lawsuit is brought as the result of a broken bat. The required amount used to be a whopping $12 million, which at $60,000 to $70,000 for a policy put many smaller bat manufacturers out of the running. Some complained it was the big boys, like Louisville Slugger, with deep pockets that instigated the insurance rule to minimize competition. In addition, each of the thirty bat manufacturers okayed by major league baseball, regardless of size, must pay a $10,000 per year certification fee to the league—another charge that has led the small companies to yell "foul."

Aluminum bats, though virtually unbreakable and capable of hitting a ball 30 feet farther, will never take the place of wood in the major or minor leagues. It would wreak havoc on the record books, requiring two sets: BA (before aluminum) and AA (after aluminum). But the main reason for the rejection is that the ball flies off an aluminum bat with such velocity that players often don't have time to react. Many high schools and colleges, after using aluminum, are now returning to the wood bat for safety reasons.

Some players have tried their own means of improving the wood bat. Sammy Sosa's corked bat, which got him ejected from the game in June 2003, was only one in a long string of bat tamperings. Albert Belle received a seven-game suspension after some corked bat shenanigans that included one of his teammates crawling through a suspended ceiling to the umpire's dressing room to covertly reconfiscate the confiscated bat. And there was Craig Nettles, who had six superballs fly out of his broken bat during a game in 1974. The hell of it is that when one does the physics, corking or superballing a bat may actually shorten the flight of a ball.

Two other types of wood bat that have recently been developed will never find their way to the MLB playing fields. The first is made of thin plies of wood, glued and rolled up like a newspaper; it's virtually unbreakable. The second is a bat made of laminated layers of bamboo, a substance with a surface hardness rivaling that of iron. Neither meets the "solid wood" stipulation in the rule book; indeed, the bamboo version—since bamboo is a grass—doesn't even meet the "wood" stipulation. And that's just fine with the MaxBat folks.

GOLF: PERSIMMON SCORES A HOLE IN ONE

If Iron Byron—the robotic golfer the United States Golf Association uses to test balls and clubs for conformity to standards—could talk, it would tell you this: the Thumper, a driver made of persimmon wood manufactured by Louisville Golf, will consistently hit a golf ball 3 feet farther than the high-tech, super-hyped, titanium Big Bertha driver. Michael Just, president of the company that makes the Thumper, will tell you something else: in this day and age when the top golfers are paid millions to play high-tech clubs manufactured by the big boys—Callaway, Titleist, TaylorMade—it's hard to spread the gospel. As with baseball bats, sports drinks, and underwear, endorsements equal big money for the players and big sales for the manufacturers—a cycle that's economically difficult for smaller businesses to break. But Michael Just, Iron Byron, and the Thumper keep on swinging away; convinced that even in this graphite-shafted, carbon-fibered, bimetal world, you can't keep a good wood club down.

Golf is one of the oldest games surviving today. In the thirteenth century, the Dutch played a game named golf wherein the goal was for the players to—much to the general populace's chagrin—hit hardwood balls toward selected targets, primarily the doors of windmills and castles about town.[6] No clubs of that vintage have survived, but written accounts reveal that the equipment—balls, clubs, "flags"—was made of wood.

The development of the club went hand in hand with that of the ball. Originally, in the thirteenth through the seventeenth centuries, balls in Europe were made of boxwood—a wood of monumental density—often turned on a lathe. Boxwood balls were prone to shattering, but also inexpensive and easy to make. It was one of the things that made the game of golf easily accessible to the common man.

Apparently, hook shots and slices were also original parts of the game. They inflicted such damage to property, life, and limb, especially when city streets were used as the course, that many laws were passed limiting the game. Though these laws put a temporary damper on the game, they did provide a written account of the game of golf as early as the fourteenth century. Records show that in England there was another problem: one of national defense. Too many men were practicing golf when they should have been practicing archery.

In the late 1600s, the feathery ball flew onto the scene and changed not only the way golf was played but also who played it. The game became one more of finesse than of brute strength. And while boxwood balls were something any fellow could whittle on his back porch at night, the making of a feathery ball was more complex. They were expensive to make, to such an extent that the game became one of privilege and wealth rather than one for the masses. Featheries were made by sewing together a spherical pouch of bull or horse hide, then turning it inside out through a very small hole so the stitching was on the inside. This was then packed tight with wet goose down. The wet leather would shrink, while the wet feathers would expand, creating a hard yet resilient ball that could sail 200 yards. Making a feathery required great skill, and even the best ball makers took two hours to create a single ball. Shank one and the stuffing might fly out, hit one in the water and it took days to dry out, lose a couple and a workingman lost a day's wages. Only the filthy rich could afford to play the game.

Featheries could withstand hits only by wood clubs, so even though iron-headed clubs had been available since the 1600s, they were rarely used. Wood reigned in the long game, short game, and putting. A play club, with a long shaft and low-loft head, was used for driving; a grassed driver had more loft and was used for long fairway shots; long-nose spoons—wood clubs with smaller heads and varying degrees of loft—were the irons of the day; a baffy spoon was a sort of pitching wedge. A putter was a putter.

Early club heads were most often made of dense woods from fruit trees, particularly apple, pear, cherry, and plum. Dogwood and hornbeam were also popular. Club heads were—as drivers and other woods are now—shaped like a hitchhiker's fist; the club head represented by the fist, and the short stub running up to the shaft represented by the thumb. Grain direction was critical. It needed to run parallel to the "thumb" in order to withstand impact, but of course, the club face would be stronger too if the grain ran parallel to *its* length. Thus, the ideal wood for a club head was one with a natural dogleg bend to it. Blackthorn trees were particularly good at yielding these V-shaped pieces. Beech was also popular, but allegedly, only if the wood was harvested from trees that grew on hills; valley wood tended to be too soft. Bow makers became the earliest club makers; they had the proper tools, skills, and—once the English and French decided to stop killing one another in droves—the time. Many blacksmiths, too, crossed over to club making. A bow saw was used to cut the head to a rough shape from a single block of wood. Rasps and files would then be used for fine tuning; clubs indeed were a custom-made affair. Connecting head to shaft was a tricky endeavor. The best club makers employed a method that was strong enough to have been used by the Romans for repairing broken ship masts. The neck of the club head (the end of the thumb) would be cut at a long tapered angle and the end of the shaft cut at a corresponding angle. The two tapers were glued together; then the joint was strengthened with tight wrappings of fisherman's twine, coated with pitch.[7]

Many early clubs had shafts of hazel—a strong wood, but one with too much "whip." Club makers turned to ash for a stiffer shaft. But around 1820, hickory—a lightweight wood, able to absorb a great amount of shock and having an almost ironlike spring to it—began to be used. If it made the supreme handle for axes, hammers, and other tools, wouldn't it also make the supreme golf club shaft? Yes. Hickory revolutionized the game. In 1922, toting the superiority of hickory, one U.S. Forest Service employee explained, "The combination of strength, toughness, and elasticity in hickory has made it the world's foremost wood for certain purposes. As a shock-resisting wood its equal has not been discovered."[8] To determine the resiliency of any particular hickory shaft, the commonly used club maker's test was to hold it loosely in the hand and bounce the butt against a concrete floor. A clear ringing sound indicated a "quick" shaft, while a dull sound portended a shaft with a slow recovery time.

Eventually, so many hickory trees were cut for clubs that demand began outstripping supply; some suppliers actually began rationing hickory to manufacturers. The average golfer had difficulty obtaining clubs with consistently good shafts. There were four grades of hickory shafts: G, O, L, and F, with G being the best and F being the worst. Almost all mass-produced clubs were in the L and F range, with the

Gs being reserved for top players. People eventually began looking at other materials for shafts.

When the gutta-percha ball—made from another material derived from a tree, but now in the form of sap from the sapodilla tree—took over in the 1850s, the game changed once more. The common man, even one with a wicked slice, could afford to play again. Since the harder balls were tougher on the wood clubs, the clubs too evolved. Club makers began securing bone, horn, and brass to club faces so they could better withstand the impact. Thick leather grips were added to help absorb the increased shock. And slowly but surely, iron club heads began replacing the long-nosed wooden spoons.

The shortage of hickory and the increased hardness of the ball led to the legalization of steel shafts in the 1920s. But steel shafts had their own repercussions on the game; they had a different "feel." Because of this, golfers became less adept at using their clubs for multiple types of shots and began carrying more clubs to compensate. The next thing they knew, they needed a bag for the clubs, and perhaps a knicker-clad lad to carry them.

Invariably, when you discuss the difference between wood and metal clubs, that word "feel" creeps into the conversation. It's a difficult concept to define, and as a golfer with the finesse of Elmer Fudd on the course, I advise you to take my opinion as a golfer with a mound of salt. However, as a carpenter who used a wood-handle hammer for fifteen years, I can relate to the sentiment. The "feel" of wood involves several things. In part it's the sound and tactile sensation wood transmits to your ears, arms, and hands. It encompasses the ergonomic and emotional aspects. It's the way a good ash or hickory handle—whether it's secured to the head of a golf club or a hammer—delivers just the right amount of power, coupled with the right amount of shock absorption. The other part of "feel" comes from the feedback your body and brain receive from each hit—a feedback that is, to some, truer with wood. It's not that you can adjust your swing in mid-hit based on this feedback (unless you can react during the 5/10,000 second the club face contacts the ball). But each whack teaches the next whack to be better. If the first whack didn't hit the sweet spot, your body learns how to make the minute adjustments so the next whack is better. For whatever reasons, wood—because it vibrates at a higher frequency than steel, because it isn't homogenous and transmits more easily read signals, because it once used to be alive—is a better teacher. Wood club enthusiasts talk about the ability to "shape shots" better with wood drivers and fairway woods.

Though metal supplanted wood when it came to shafts and irons, wood, particularly persimmon, clung to its grip in one area: drivers and fairway woods. As recently as 1984, woods actually made of wood were used by nearly 90 percent of touring professionals. Indeed, during the 1970s persimmon was being turned into club heads at such a rapid pace that the major suppliers rationed it, as with the hickory of old. But even that stronghold eventually slipped away. Nonwood woods were thrust into the spotlight in 1991 when John Daly—a twenty-five-year-old unknown pro and ninth alternate to the PGA Championship—used a titanium-shafted Kevlar driver to outdistance the competition on his way to winning the tournament. He averaged 303 yards per drive—not bad for a fellow who didn't have the opportunity to shoot a practice round.

In 1997, Justin Leonard and Davis Love III became the last well-known PGA players to trade in their wood woods for those of metal.[9] As late as 2001, Lousiville Golf was custom building wooden woods for Bob Estes, the last holdout on the tour to use them. But then they got the call: Cleveland Golf had offered Estes, a rising star, a substantial amount of money to play with their clubs. He could no longer afford to play with wood. Today it's estimated as that few as one-tenth of one percent of all golfers swing a wood made of wood.[10]

But Louisville Golf is a survivor. Elmore Just, Michael's brother, founder of the company, and "high priest of persimmon" forged ahead under the mantra "Persimmon is nature's gift to club makers and golfers." Persimmon—as dense and strong as other members of the ebony family to which it belongs—is the near-perfect club head wood. In his book *The Persimmon Story,* Just explains, "Persimmon is both a heavy and a porous wood. This complex structure results in a unique material that is not dense and dead like lead, but hard and resilient. It is literally alive." The head of a driver optimally weighs 7 ounces, and a piece of persimmon of that size weighs—ever so conveniently—7 ounces. Once seasoned, persimmon is nearly immortal; it's impact resistant, extremely hard relative to its weight, and almost impossible to wear out. Like a cue ball, it absorbs very little shock energy and quickly returns to its original shape. "It makes an excellent hammer," says Michael Just. "There are harder woods, but they're brittle; there are denser woods, but they're too heavy; persimmon has a grain structure that makes it perfect."

In standardized hardness tests where the pressure required to sink a 1/2-inch steel ball halfway into a board is measured, sugar maple and white oak—both woods renowned for their hardness—require only two-thirds the force required by persimmon. Weaving shuttles made from persimmon can last a thousand hours or more, where other woods can't take the beating for an hour. It's one tough wood.

And then there are the aesthetics. Persimmon all but glows when polished. According to Just, "Persimmon drivers appeal to the same person who likes leather seats in their car or a desk made of wood." Crafting a wood driver requires more than a hundred different operations and six weeks: four times as long as it takes to churn out its metallic counterpart. The process starts with drying hand-selected flitches in a radiofrequency vacuum dryer. Club heads are then rough shaped using a master pattern and tracing lathe, a gizmo similar to a key-duplicating machine. The heads are then sanded, and holes are bored to accept the metal sole plates and faces. Installation of the shaft, more sanding, detailing, painting, and clear coating follow.

Through this process, it's much easier to customize a wood driver made by hand, compared with a metal driver cranked out by machine. Bob Burns, a club maker from Wisconsin, explains, "Every day I tell people that I can bore a wood club for them that will allow them to hit the ball straighter and more consistently, and some of those people listen to me and are the happiest people in the world. There's so much more you can do with wood that you can't with a metal club

A worker stands in front of the tracing lathe used to rough-turn persimmon wood club heads. The master shape is inserted in the top spindle to guide the cutters as they shape three heads in the spindles below.

to fit it to a golfer, such as making it more upright for a taller person, or closing the face a few degrees for someone who fades."[11]

The recent increase in orders at Louisville Clubs is due in part to performance, part to nostalgia. For $489, their Thumper persimmon driver, oversized and complete with cocobolo accents, can be in your golf bag. And not only do you get to carry a work of art around the course, you get to hear the solid whack of wood instead of a hollow metallic ping when you tee off. If you're a true wood-loving golfer, you can provide company for your driver with a Louisville putter crafted from walnut, dogwood, cherry, or good old hickory.

Still, for almost all golfers today, the only wood object in their bag is a 2-inch-long tee. That's a radical shift from days gone by, when the sand mold—a metallic, funnel-shaped doodad that golfers would pack with dirt, then turn upside down to create an anthill to tee off from—was the only thing that *wasn't* wood. But what about that teeny minority?

They belong to a subculture that plays hickory golf. They approach their pastime with all the zeal of Civil War reenactors and muzzle loaders. They use restored clubs from the 1900s with hickory shafts and woods with real wood heads. The dress code for the 2006 National Hickory Championship dictated that men wear "bow tie, neck tie, or cravat," and upon their heads "wide-brimmed straw or felt hat, top hat, bowler or derby or flat cloth cap. No baseball-type caps or visors." For ladies, "approximately ankle length" skirts or dresses are required. For those playing in the Nineteenth Century Division, clubs are to be carried under the arm, without a bag.

Feel like taking a swing at it? Play Hickory is a company specializing in supplying vintage clubs for special events. Rent any of their eighty sets of clubs and you'll get a driver, four irons, a putter, and a small bag—just like the golfers of yore. You won't get an 1840s Peter McEwan–built long-nose play club worth $6,500, or a mint-condition feathery ball that can bring upwards of $40,000 at auction, but you'll get some cool vintage stuff to tote about for a round or two.

And will wood golf clubs remain the province of only a dedicated few? Probably. But those dedicated few will have an extra dimension to their game: the mystery and history of wood. As Elmore Just presciently explained over twenty years ago, "Like Bobby Jones and his Calamity Jane putter, there is a small detail that makes my club personal and work for me. The same is true for most serious golfers. A day may come where all golf woods will be made by some new process. They will have their advantages and disadvantages. But if that day comes I still want to be out there competing and playing with one of the last hand-crafted custom-fitted persimmon golf clubs."[12]

TOSSING TELEPHONE POLES AND OTHER CURIOUS SPORTS

If you ask a Scotsman about the finer nuances of the caber toss, you might hear something like this: "First you get really drunk and then you throw a telephone pole." If you doubt the veracity of this, you might hear "I knew ya wudent undastand."[13]

The caber toss of Scotland reduces wood sport to the very basics. The goal is to vertically balance a 16- to 20-foot pole weighing 80 to 180 pounds in one's cupped hands, then run and throw it end over end, so the bottom lands directly opposite the thrower. If the thrower is at 6:00, the judges want the caber's former bottom end to land at 12:00. A caber pointing at 1 o'clock is better than one landing at 2. All of this with a kilt on.

The origins of the sport are unknown. Some surmise that the

A kilt-clad contestant takes a running start during a caber toss competition. The cabers can measure up to 20 feet in length and weigh up to 180 pounds.

caber was originally a weapon tossed over enemy fort walls. Others think that it was used in forestry as a way of moving felled trees to a more convenient spot. The most often-cited explanation is that it was a way of quickly creating a makeshift bridge to cross small streams.

If you wish to throw a caber without donning a kilt or chancing a hernia, you can do so in virtual space at Electric Scotland.[14] Be forewarned—if you don't throw it far enough, a digital Scot will tell you to "Eat more porridge."

Pole vaulting is another sport that originally used wood in its rawest form. Track-and-field historians maintain that the sport may have originated in the Netherlands, where people used long poles to vault over wide canals rather than wear out their clogs walking miles to the nearest bridge. Others maintain that pole vaulting was used in warfare to vault over fortress walls during battle. In these first two suppositions, pole vaulting shares much with the caber. The activity became a competitive sport in the late 1700s, and as in many sports, the bar was raised (literally) as the equipment improved.

Initially, vaulters used poles of wood—most often ash, hickory, and hazel. Sheer strength, jumping ability, and speed determined the height a vaulter could attain, since there was very little bend and snap to the poles; the heights attained topped out at about 11 feet. In the 1800s, vaulters transitioned to bamboo, a material with more spring that pushed heights to around 13 feet. Now, with the use of carbon fiber poles—poles that can bend upwards of 110 degrees—vaulters routinely reach heights of 19 feet. But heights like these come at a cost. Between 1983 and 2001, fifteen high school athletes died while pole vaulting.[15] Perhaps that 11-foot wood height is a good height to stick to.

Another participant in the world of unadorned wood used for sport is the boomerang. Some surmise that the classic V-shaped item was developed by a nearsighted hunter who tired of walking great distances to pick up his errant weapon. But in all probability the boomerang was designed more for sport and play.

The early throwing wood—a lopsided affair, heavily weighted on one end, meant for throwing, bonking, and killing but not returning—was used with regularity by early hunters to kill hares, birds, and even fish. It is most often associated with Australia. But its use in Africa can be traced back to 6000 BC, and rock paintings in Europe show throwing woods in use as early as 2000 BC. It's estimated that Native Americans began using them two thousand years ago. But these throwing woods are not boomerangs, the clear distinction being found in the concise words of Tony Butz, founder of the Boomerang Throwing Association: "If it doesn't come back, it's not a boomerang."

The boomerang that most of us think of when we hear the word "boomerang" was a baffling thing to the Europeans arriving in Australia. When Captain James Cook first saw aborigines with boomerangs tucked into their belts in Botany Bay in 1770, he mistook them to be wooden swords. His botanist, Sir Joseph Banks, likewise thought them to be a type of Arabian scimitar.[16] Some Europeans, combining stories related to the throwing wood with those of the returning boomerang, reported that the aborigines could toss a boomerang, kill a kangaroo, and catch the weapon upon its return—an aerodynamic impossibility.

The original shape of a traditional boomerang is based on both aerodynamics and the natural shape of tree roots. The boomerang is strongest when the grain of the wood runs along its entire length. Tree roots often yield continuously curved pieces of wood, and the natural 110-degree bent is used for traditional boomerangs. Today's boomerangs can be made of aluminum, carbon fiber, or Lexan, but their roots are in wood. Boomerang traditionalists use only those made of wood, and scads of gorgeous, aerodynamic, high-flying boomerangs of solid wood and plywood fly today. There are boomerang clubs worldwide, including the Mothers Against Boomerangs student organization at the University of Texas, which encourages all interested parties to BYOB-IYGO (Bring Your Own Boomerang—If You've Got One).

A World Cup competition is held every two years. Official boomerang competitions have fostered record tosses of 780 feet, record consecutive catches numbering 1,251, and a maximum time aloft record of 179.94 seconds (though most of these records are held by nonwood boomerangs). Immortalized in boomerang history is John Gorski's 1993 noncompetition toss in which his boomerang stayed aloft for over 17 minutes and reached an estimated height of 650 feet after hitting a thermal draft. He made the catch 230 feet from the launching point. Captain Cook would have been impressed.

THE ART OF THE POOL CUE

From the moment Vincent Lauria breathlessly and ever-so-reverently asked "Is that a Balabushka?" when Fast Eddie Felson presented him with a gorgeous pool cue, the value of a cue made by the late George Balabushka has soared. The fact that this exchange took place in the movie *The Color of Money* and the roles were played by Tom Cruise and Paul Newman may have helped a bit. Today a Balabushka cue—with good provenance, since so many forgeries are in circulation—will set you back $10,000 to $20,000. Are they worth it?

Not according to cue maker Arnot Q Wadsworth, III. "To be honest with you [Balabushkas] are horrible cues. I've never even talked to anyone who liked the way they played. And I know a hundred cue makers that make a better cue than Balabushka." Arnot should know; he's one of them.

Arnot just *looks* like a guy you'd find sauntering through the world of pool cues and pool halls. By appearances he could be the kid brother of Minnesota Fats. And while Arnot shoots a wicked game of pool, he makes an even more wicked cue—a cue that doesn't come cheap. A 20-ounce Diminishing Diamonds pool cue from Arnot Q Custom Cues, complete with 53 ivory inlays, ebony accents, and maple shaft will set you back $3,299. Without upgrades. He's made others costing up to $6,000 and knows of at least one custom cue that sold for $100,000. His own personal cue is made of cocobolo and bird's-eye maple and contains inlays consisting of 296 pieces of sycamore, holly, and bloodwood. Each Arnot cue is a work of art.

Arnot Wadsworth at work in his pool cue workshop. Some of his cues contain three hundred individual pieces of wood and sell for $6,000. "But we're not going to do anything to the cue that's not good for the cue," he explains.

So what's the difference between a $3,299 cue and one costing $32.99? Absolute straightness, for starters. "Wood in its natural state has a memory," explains Arnot. "It remembers where it came from, and it remembers how it grew. What we want to do is kill that memory if it's a bad memory, and keep it if it's a good one." Arnot is referring to the process he uses to season and incrementally reduce the size of the wood in order for it to properly stabilize. It takes years. Arnot starts with a piece of wood that's been seasoned three to five years, rips it into 2-inch by 2-inch blanks, allows it to rest for another year, trims the edges so the blank is octagon-shaped, lets it season for another year, turns it into cylindrical shape, allows it to season for another year, and finally (finally) turns a cue.

Professional two-piece cues—the type that screw together—consist of the shaft and the butt. The shaft, the narrow half holding the tip, is made from a single piece of wood, but the butt—at least Arnot's butt—is made from three pieces; the forearm (usually 13 inches long), the handle (usually 12 inches long), and the butt sleeve (the thick end of the cue, usually around 4 inches long.) At first glance this seems odd, as if three separate pieces might increase the possibility of breakage or warping, but not so.

"With three short pieces, even if one warps a little, it won't be as pronounced as a warped single-piece butt. Any movement is interrupted by the splice," explains Arnot. The three pieces are joined with indestructible fiberglass rods and epoxy glue. Absolute straightness.

In the end, all the pieces must be joined in such a way that the cue performs like a single length of solid wood. Arnot uses inlays and veneers, and he utilizes a CNC milling machine for some tasks, but he says, "We're not going to do anything to the cue that's not good for the cue. Every time you take a shot it's like hitting your cue with an 11-ounce hammer [the weight of a cue ball]. You do this thousands and thousands of times, and unless you've built it right, things will shake loose." If you want to learn how to build your own masterpiece, you're in luck. For four grand you can become a student in Arnot's two-week Cue Making School.

Nearly any dense hardwood can be used for cues. In Arnot's shop you can find ziricote, Gaboon ebony, bird's-eye maple, camateo, purpleheart, bloodwood, and pink ivory; an exotic wood from Africa said to be rarer than diamonds. "My wife says, 'you're sixty-two and you're only going to turn cues for another seven or eight years, if you live that long. And you've got enough wood in there to build five thousand cues. What's wrong with you? Do the numbers!' But the wood blanks are my babies. I love 'em. They're a treasure." Like Arnot.

TENNIS: THE RACKET ABOUT WOOD RACQUETS

In "The Feel of Wood," Marshall Fisher describes his experience playing tennis in the Woody Tournament of Cape Cod: an event where the racquet rules specify, "No steel, aluminum, graphite, titanium, or composite need apply. If it didn't come from a tree, leave it at home."[17]

Fisher describes the extra skill required to make and place each shot, the techniques slumbering deep inside muscles that were awakened, the need to actually volley versus simply sticking out a high-tech racquet, and watching the ball bounce off to score. Most important, the game was more fun.

Fisher isn't alone in his sentiments. Many players and fans alike feel that new generations of racquets have not only changed the game but delivered it a kill shot. Racquets, they maintain, should have gone the same route as baseball bats. Let amateurs play with the space-age composites and metals, but have the pros stick with wood, where the trajectory and speed of the ball are determined more by strength, finesse, and skill than by high-tech equipment. But, alas, it's too late: Howard Head—the founder of Prince racquets, the man who took what he'd learned in developing fiberglass skis and applied it to tennis, the developer of the oversize composite racquet—started the stampede twenty-five years ago, and there's no leading the horses back to the stable now.

Tennis has roots dating back to the days of the ancient Egyptians and Moors, when ball games were played as parts of religious ceremonies. The very earliest racquet was the human hand. As hands became sorer and the game more advanced, leather playing gloves were developed, followed by gloves with webbing between the fingers. The first real racquet was created when someone attached one of these gloves to a wooden handle. As the game evolved and spread to Europe during the sixteenth through the eighteenth centuries, early French players would signal the start of the game by yelling "tenez," meaning, essentially, "Comin' at ya!"

Initially, the game headed more in the direction of racquetball, with solid balls made of wads of hair, wool, or cork wrapped in string, cloth, or leather. But by the nineteenth century, a product of the para rubber tree coupled with a process called vulcanization (thank you, Charlie Goodyear) revolutionized the game. It produced a ball that was soft enough not to damage the courts made of grass, yet resilient enough to stand the endless smashing of the racquet. The game continued to evolve. Organized clubs formed. In 1875, the All England Croquet and Lawn Tennis Club needed to raise additional funds to cover the rent on their 4-acre site and started one of the earliest tournaments. Their site was in a London suburb by the name of Wimbledon.

Wood created the framework for the original racquets used in squash, tennis, badminton, and racquetball. At first, most racquets were single loops of ash or some other wood amenable to bending; the loops were bent to shape and secured with animal glue. Frames were strung with animal gut; that of the sheep preferred. These single-layer frames were strong but, under the tension of strings, very prone to warping.

Eventually, thinner layers of veneer, bonded with glue, were used to create racquets that were stronger, more consistently shaped, and less prone to warp. They were crafted from three or four woods, each with its own special qualities. Multiple strips of ash or beech—strong yet flexible—were steam-soaked so they could be easily bent around a form to create the main frame. Synthetic resin glue was used, and heat was applied to speed the curing time. One manufacturing process involved gluing up and bending wide strips of veneer to create a racquet the thickness of three normal racquets. These blanks were then bandsawn lengthwise to create three individual racquets. Hickory would be applied as an outer rim layer because of its superior wear resistance. Strips of sycamore and mahogany were used to build up the thickness of the handle. Lightweight mahogany was often used for the throat; the triangular section where the head transitions into the handle, and obeche was used in the shaft as a lightweight filler.[18] After shaping and sanding, the racquets were placed in a multi-spindled drilling device that bored thirty holes at once. Each blank was run through twice, and four additional holes were hand drilled to create the sixty-four holes required for the strings.

Amazingly, the wood racquet underwent few changes from the early days of Wimbledon up to the 1960s. The Maxply—a no-frills au naturel wood racquet—was first sold in 1931 and proceeded to become the most popular in the world. Eventually, some manufacturers experimented with laminating wood to other materials. Then, in 1967, Wilson developed the aluminum racquet. The racquets were lighter, the frames were thinner, and the sweet spot was sweeter. As racquets evolved, the string area also grew, eventually becoming 50 percent larger than that of the old 65-square-inch wood versions. And while wood racquets weighed as much as 14 ounces, those made of the new composites dropped in weight to as little as 7 ounces. Aluminum led to graphite, graphite to titanium, titanium to tongue-twisting materials like Yonex's 0.7-nanometer fullerene carbon fiber.

The last wooden-racquet duel at Wimbledon took place in 1980 between Bjorn Borg and John McEnroe. Some consider it the greatest game ever played. In 1981, Bjorn Borg was the last male player to win a major tournament—the French Open—using a wooden racquet. In 1982, Chris Evert won the U.S. Open with a wood racquet. And in 1983, Bill Scanlon played the "golden set," the only perfect set ever played in professional tennis—twenty-four consecutive points—using a wood racquet. But 1982 was the turning point; more oversize, nonwood racquets were sold than woodies that year.

As with golf, there is the constant mention of the difference in "feel" between wood and composite racquets. The latter delivered more speed, but the former offered better control and touch, and a certain connectedness to both the roots of the game and the feel of the ball. Bjorn Borg surely felt that way. When he emerged from a nine-year hiatus from the pro circuit in 1991, he did so still grasping a wood racquet. Fisher describes the scene: "He looked like one of King Arthur's knights on a Connecticut Yankee's backyard court. Young powerful paladins battled each other on the red clay of Monte Carlo, blasting serves with the latest generation of oversized, wide-body racquets. And there was Borg, stepping on to the clay, pigeon-toed as ever, dangling from his right hand a black wooden anachronism. . . . He never had a chance."[19]

The cry in the wilderness, sounded by former tennis greats like McEnroe, Boris Becker, and Martina Navratilova, is that the return to wood would slow down the game, make players move around more, and increase the number of touch shots; this instead of players staying anchored at the baseline, crushing the ball back and forth in long, drawn-out rallies. There's more control, better feel, less stress, lower incidence of tennis elbow, fewer shoulder injuries, and longer careers with a wood racquet.[20]

The Wilson Autograph Jack Kramer is, to many, the standard by which all other wood racquets are compared. It was used by McEnroe, Tracy Austin, Arthur Ashe, Chris Evert, and other greats to win more major tournaments than any other racquet in history. Margaret Smith Court, high priestess of women's tennis, declared it "the greatest wooden racquet ever made." And Pete Sampras—winner of a record fourteen Grand Slam titles—maintains that when his son is old enough to play, he'll start out using a wood Jack Kramer.

If you're a wood tennis racquet lover, don't despair. There are wood racquet tournaments aplenty. You can find NOS (new old stock) Wilson Autograph Jack Kramer racquets that "have never been strung or swung in play" at www.retroracquets.com. The company acquires vintage racquets by scouring the basements and back closets of "going out of business" sporting good stores. Cost? $150, complete with original tag.

But there's no turning back the hands of time. Wood is down, love–40. One new racquet has a computer chip built into the handle that allows the racquet to stiffen upon impact with the ball, leading McEnroe to say, "I think the sport has lost something. It's lost some subtlety, some strategy, some of the nuance."[21]

LUMBER JACKS AND LUMBER JILLS

Wood can be used to make sporting equipment, but it can also be used as a sport in and of itself. Nowhere is this more evident than in the world of lumberjack competitions.[22] And in this strange competitive world, perhaps the strangest contest of all is the open modified chainsaw event—the event those in the know simply call the hot saw. The goal is to cut three thin slabs from an 18-inch-diameter white pine log within 6 inches of the end. Cut beyond the 6-inch ring marked around the log, and you're disqualified. It's down-up-down as fast as possible. What seems implausible are the machines the participants are holding. The engines are modified snowmobile and motocross engines, with 50 to 60 horsepower, running anywhere from 10,000 to 12,000 rpms. The saws have 24-inch bars shaped like ironing boards and an average weight of 65 pounds.

The only three regulations are these: (1) they're single cylinder, (2) they hold their own fuel and oil, and (3) they're pull started. These unworldly-looking monsters are—without exaggeration—the loudest thing I've ever heard. In order to keep their weight down and their power up, mufflers, safety devices, and other niceties have been dismissed. The participants are allowed one minute to warm up their saws, after which they turn them off and place both hands on the log in front of them. At the signal, they reach down, start their

A competitor in the hot saw competition cuts the first of three slabs from an 18-inch white pine log. The winning time—under six seconds—included starting the saw while on the ground.

saws, and make the three cuts. It's all over in as little as 5.93 seconds. NASCAR—meet your match.

I'm at the Midwest Lumberjack Championships. John Hughes, the Mad Kiwi, who serves as both manager and announcer, refers to the events as sports and the participants as athletes—but these are sports and athletes unlike any other. Foot speed is not a factor, vertical leap is inconsequential, endurance comes into play for only a matter of seconds, and nothing needs to get caught or hit with a stick. What is required is raw power, technique, and balls—even if you're a woman. Some of the participants fit the stereotypical image of the young buff lumberjack—but good luck trying to pigeonhole the rest. One father-and-son team in the two-man sawing competition easily go three bills apiece; one of the top women contenders wouldn't tip the scales at 100 pounds soaking wet. The oldest participant is seventy-one, and the youngest is fifteen. Pant waist sizes range from 28 to the mid-50s.

If you think physiques are all over the map, look at the occupations. Almost no one—even the top competitors in the world—makes a living solely via prize money. Some stay close to the sport through their vocation; there are chainsaw wood carvers, sawmill managers, tree loppers, and lumberjack entertainers. But you'll also find chicken farmers, police officers, landscape designers, and one participant who lists himself as a massage practitioner/skidder operator. The women's Midwest champion, Nancy Zalewski, is a thirty-seven-year-old chemist with a powerful build. Her strongest rival, Linsday Daun, holder of two world records in the single sawing event, is a sophomore at the University of Wisconsin/Lacrosse and half Zalewski's age. She looks more like a figure skater than a lumberjill, but don't mess with her.

There are two nonhuman components to the sport: the wood that gets cut and the tools that do the cutting. Both are remarkable works of precision. The Mad Kiwi has been at the site for three days preparing the wood; white pine is used for the sawing competitions and aspen for the chopping. For the playing field to be as level as possible, the wood needs to be as consistent as possible. "And it's a bloody lot of hard work," explains Hughes, a New Zealand native. "We brought in nine and a half cords of 40-foot logs, unloaded them by hand, and sorted out the best logs." Since consistency of the wood is critical, all the wood for a single event comes from a single tree. The 2- to 3-foot lengths are labeled as they're cut to length. There's yet another selection process before each section is placed on a gigantic lathe, which turns them down to an exact diameter. Diameters range from 18 inches for the hot saw competition to 11 inches for the women's underhand chop. If a visible knot is discovered in a log, it goes into the firewood pile; in a sport that's timed in thousandths of a second, even the smallest flaw can make the difference between first place and eighth place. Knots are encountered in competition, but at that point Hughes and crew have done everything they can to sort them out.

Since green wood cuts and chops the best, great care is taken to keep the log sections at a high moisture content. Each white pine segment is wrapped in plastic immediately after being cut, stood on a tarp covered with hay, sprinkled with water, and covered with more hay and another tarp.

Preparations are a bloody lot of hard work, and so are the competitions. With the exception of the axe throw, the goal of all the competitions is to chop through or saw through a log as quickly as possible. The underhand chop involves balancing on a cradled 3-foot section of aspen—11-inch diameter for women, 13-inch for men—while chopping through it. Since the first blow is so powerful that it sometimes shears off the entire face of the log, there's a slabbing rule: four nails must be driven near the end of the log opposite the angle of the first blow to keep the log—and the chopper's foot—intact.

The mission in the standing block chop is to whack through a vertical 12-inch aspen log. If you're Jason Wynard, it takes you 12.8 seconds.

Felling logs by sawing was favored by the old-time lumber barons; the thin kerf taken by a saw meant that more of their tree made it to the mill, versus winding up as woodchips generated by an axe. People setting up for sawing competition—whether it's singles, two-man, Jack and Jill, or Jill and Jill—have exacting routines. A wide stance, usually twice as wide as the shoulders, is critical, and competitors drive blocks into the platform with the precision of a track-and-field sprinter to establish optimal foot position. Sawyers are allowed to cut a preliminary half-inch-deep kerf—one that's carefully measured by judges—into the top of the log as a starting point. The fastest route through a log is a straight line, and when I measure several of the 18-inch-diameter rounds cut after one competition, their thicknesses don't vary more than 1/8 inch anywhere along their perimeter.

The type of wood used varies from country to country, and times also vary accordingly. Zalewski, a world record holder in crosscut who competes worldwide, explains, "In Australia and New Zealand they use really hard silver top ash for chopping. It can take a minute to chop through that, while it only takes half that to chop through the same size aspen log in the States."

It's an exceedingly expensive sport. A good two-man crosscut saw costs $2,000. A top-notch axe starts at $500. A good hot saw will set you back five grand. Travel adds to the expense. You need different saws and axes, sharpened specifically for cutting through softwood and hardwoods. One male contender travels with sixteen axes and has another forty at home. Another competitor has thirty saws hanging on his basement wall.

The sport is not without hazard to both participant and equipment. A glancing blow off a log during an underhand chop in a previous competition took the entire calf muscle of one female contestant. Another contestant destroyed every tooth of a $1,500 saw when he hit a whorl of nails embedded in the log, remnants from when the tree was used long ago as a fencepost. But given the sharpness of the tools, the speed at which they're used, and the fact that some contestants favor flip-flops and tennis shoes, the sport has surprisingly few accidents. Still, throughout the competition, the ambulance and paramedics are stationed less than 100 feet away.

Only three crosscut saw makers are left in the world today, and Jean-Pierre Mercier, sitting across the picnic table from me—who's also a top-notch competitor—is, according to most, the best. As he explains the process you understand why he's earned the reputation. He starts

Lindsay Daun, youngest-ever world champion in the women's single buck-sawing event, chats with her father, Dennis Daun, while warming up for the competition. Dennis took third place in the two-man sawing event.

with 500-foot rolls of steel, the same material used for making the mammoth bandsaw blades in lumber mills. He uses an AutoCAD program to map out the length, gullet depth, pattern, and angle of each tooth. Next a computer-guided laser cuts out each saw blank. From there, J.P. completes each saw—one tooth at a time—by hand. He adjusts the set, the angle each tooth angles outward, at between 15/1000 and 19/1000 inch depending on the saw's intended use, then sharpens each tooth with precision. When I ask him about his training and background, he responds "trial and error," but it's more than that. His father was a logger, and his grandfather worked in the logging camps when crosscut saws were still used.

When I ask him how a lumberjack from the old days might fare in today's competitions, he responds that there isn't a standard chainsaw that can cut through an 18-inch round of white pine as fast as a crosscut saw. When I look into the record books I see he's right. The fastest time for a stock chainsaw is 15.4 seconds; for a one-man cross-cut saw, it's 10.3.

Lumberjacks—whether they trudged through the forest a hundred years ago with a crosscut saw slung over their shoulder or gallop through the woods today on a computer-aided harvesting machine—have always provided us with one essential item: the lumber for building our houses. We use it for the roof above us, the floors below us, and the walls in between.

Next we'll look at some of the structures wood has built.

CHAPTER 6

Wood as Shelter

Years ago, I attended a seminar conducted by Joseph Lstribruk of the Building Science Institute of Canada. At one point during his lecture, he paused and said something to this effect: "Wood. It shrinks. It expands. It twists. It bows. Its strength is unpredictable. Insects eat it. It rots. It burns. If an inventor were to present this product to a panel of building inspectors today, they'd ban it on the spot." Wood certainly does have its flaws; yet, when it comes to shelter it certainly has its strengths.

We can use wood for everything from assembling rustic pine log cabins to crafting polished walnut-paneled mansions. And in the most basic of all basic forms of shelters, trees themselves can be lived in. Now that's an interesting place to start.

LIVING IN TREES: FROM PAPUA, NEW GUINEA, TO WASHINGTON STATE

Tree houses come in all shapes, sizes, and heights. And they're used for a variety of motives. Witness the story of William "Scotty" Scurlock, who began building a tree house in a cluster of seven fir trees near Olympia, Washington, in 1986—and continued building for the next ten years. By the time he stopped, he'd created a 1,500-square-foot three-story treetop paradise, with "ground level" starting 30 feet up. The tree house contained plumbing, electricity, thirty windows, a wood stove, a firehouse pole for fast descents, and a sundeck with a shower perched at 75 feet. Unfortunately, the tree house also contained fake mustaches, semiautomatic weapons, and $20,000 in cash. The reason he stopped building was death. After his seventeenth heist—his last—which netted $1 million, the law caught up with him.[1]

Half a world away, you can find the tree-dwelling Korowai and Kombai people of the deep jungles of Indonesia. When there's conflict with rival clans or a stronger need to make it difficult for sorcerers to access their dwellings, these people build their houses at heights of up to 150 feet in the trees. When tensions are less, their homes may be only 15 to 30 feet up. They construct their dwellings by sawing off the top of a sturdy tree or group of trees, then, working from scaffolding, they craft a floor framework of rattan and cover it with bark and leaves from surrounding trees. Portions of the floor are supported by additional stilts extending to the ground. The roof and walls are made of sago palm fronds. Fires are built on a latticework of mud-covered rattan set over a hole in the floor; if a fire gets out of control, the area can be cut away quickly.

The Kombai and Korowai are not the only people to live in tree houses. In 1829, while in South Africa, British missionary Robert Moffatt discovered an enormous ficus tree containing seventeen aerial abodes occupied by families. They had created their unique dwellings in order to protect themselves from the lions that abounded in the area. When tree enthusiast Thomas Parkenham went in search of the tree in 1999, he thought he'd found this legendary tree, only to discover that the tree dwelling he discovered had been left by a television crew after filming a version of *Robinson Crusoe*.[2]

Tree houses can be simple or elaborate, blueprinted or seat of the pants, cheap or expensive. One tree house in Florida cost more than $200,000 to build. The tree house I built in the giant maple in our backyard at the age of eight cost little beyond the sacrifice of my great-grandmother's walnut dining room table leaves used for the floor.

A tree house of the truest sort in Indonesia, built by members of the Kombai tribe. The more at odds they are with the spirits and each other, the higher the house.

Since any dwelling is only as solid as its foundation, selecting the right tree is important. In the deciduous family, oaks, maples, and beech all have the classic branch configuration for comfortably accommodating a dwelling. Sugar maples—tapped with metal pipes on a regular basis for their syrup—don't give a second thought to having a few extra bolts being embedded in them. In the world of conifers, pine, fir, and hemlock are best used in groups. Redwood stumps make excellent home perches. Many redwoods of yore were so massive that they were cut off 10 to 15 feet above ground level simply because early saws weren't long enough to get through them farther down. A 10-foot-diameter, naturally rot-resistant stump makes an unbeatable foundation.

Trees to avoid include cottonwoods, which are massive but are subject to disease and have a soft, spongy wood; paper birches, which are short lived, usually growing to 50 feet, then keeling over; and poison sumac, which produces a sap that causes a skin rash.

But large or small, making a solid yet noninjurious attachment to the tree is critical. Contrary to popular belief, trees can easily accommodate a bolt or two. Scott Baker, an arborist from Seattle, says, "Any arborist that's been around for a while has seen bicycles enveloped by a tree with no obvious problem. They're very adaptable. So we know that you can put hardware in a tree with minimal long-term effect, but you have to do it correctly."[3]

If you wish to sample life in a tree without going through the rigors of building a tree house, you may want to schedule your next vacation at the Out 'n' About "Treesort" near Cave Junction, Oregon. There your choice of accommodations includes the 35-foot-high Forestree, which sleeps four and includes a hoist for your luggage; the TreePee, which is exactly what the name implies; the Peacock Perch ($120 for two per night), which includes a refrigerator and a chamber pot; or The Suite, which includes a deluxe full bathroom, kitchenette, and heat, all of which will set you back $200 per night.[4]

A nonprofit group called Forever Young Treehouses has a goal of building at least one handicapped-accessible tree house in every state

Baobab trees have been used for houses, bars, water towers, churches, burial crypts, and—in the case of this tree at Kayila Lodge in Zambia—bathrooms.

by 2008. Founder Bill Allen's philosophy is that any person who can breathe can get into one of his tree houses. Most of the tree houses built so far are 12 to 15 feet off the ground and large enough to hold eight to ten kids in wheelchairs. Access ramps are often over 140 feet long in order to remain at an angle that wheelchairs can easily negotiate. Allen contends, "It's pretty hard to be depressed or sad or angry if you're just sitting on a tree limb." Other benefits of tree houses include "reversal of aging, hilarity, stress reduction, insomnia relief, your neighbors will be envious, and you'll eat cookies and drink milk."[5]

Then there are those who live in trees on a more literal basis. The tree trunk that indisputably offers the best architecture for natural housing is the portly baobab (only the giant sequoia exceeds it in girth). These trees, with often bare, rootlike branches that make the tree look as if it's growing upside down, can be found in the remote deserts of Australia, Africa, India, and Madagascar. It has adapted to the dry climates by evolving a rotund bottlelike trunk, which can store copious amounts of water. Man has taken advantage of this natural design for purposes of shelter and storage. The trunks, when hollowed out, have been used for centuries for pubs and prisons, housing and storage barns, toilet stalls, bus stops, and churches. Some baobabs live two thousand years or more.

One baobab in South Africa was christened the Murchison Club and was used for years as a bar frequented by prospectors and miners during the nineteenth-century gold rush. The Platland Baobab also in South Africa houses another 35-foot-diameter pub. Yet another baobab in Namibia was "remodeled" into a toilet, complete with flushing system. If you're a tree worshipper you can commune with nature inside a giant baobab—in St. Mary's Church near Keren, Eritrea. People utilize the baobab even in death. In certain parts of Africa, the belief is held that musicians and poets are possessed by the devil and their bodies will pollute the earth if given a standard burial; the solution is to entomb them in a baobab.[6]

If you'd like to live the baobab life for a stint, head to Tarangire Treetops Lodge just outside of Tarangire National Park. You can spend a night or two in one of twenty luxurious baobab treetop suites complete with flushing toilet, hardwood flooring, deck, and spiral staircase. Note: your accommodations are atop the tree, not within it.

The baobab can also serve as a dandy water tower. In one case, a tree was hollowed out to catch rainwater, reportedly stashing a cache of up to 1,200 gallons. A hole was drilled and a plug inserted to create a tap that people could utilize to fill other containers. But not just the trunk serves as a source of water. The roots can be tapped, and the clefts in large branches also act as minireservoirs for humans and animals alike to drink from. Even the leaves can be chewed in times of drought.

Individual trees of other species have challenged the housing capacity of the baobab. The hollow trunk of the Bowthorpe Oak in Lincolnshire, England, is spacious enough to have once accommodated twenty dinner guests of the squire of Bowthorpe Park. All very interesting. But the 99.99 percent of us who do not live within range of a baobab tree have had to find other ways to utilize wood to create shelter.

THE HISTORY OF HOUSING FROM LOG CABIN TO, WELL, LOG CABIN

When we think about the homes of the earliest pioneers, we often conjure up romantic images of children playing checkers on a wood floor, Ma stitching a comfortable quilt, and the family huddled around the warm glow of a fireplace. The truth of the matter is that there was no wood floor to play checkers on, the quilt was being made because the cabin was as leaky as a sieve, and the family was huddled because the dwelling was so small there was no other choice. Early housing was a matter of survival. It went up quickly and was focused on functionality, not romance.

Housing in the New World took off in two basic directions. The Scandinavians who settled along the Delaware River and other areas opted primarily for rough log construction. Clearly, they knew how to build well; one log home built in Norway in 1250 is still standing (though it's now standing in a museum). As these poineers moved ever westward, they shared the skills and techniques needed to build them. Trees were plentiful, the skills were well-entrenched, the tools needed were few, and their houses—however crude—went up quickly, though many were little more than windbreaks.

When the English and others arrived in America, they too often continued to cling to their notion of what a house should be—and that notion was of the timber frame houses in which they'd lived in Europe. They had to have flat walls; log construction was strictly for barns and corncribs. Timber frame houses were slow, labor intensive, and tedious to build. Timbers had to be squared off with a whipsaw or adze, then be connected with specially cut joints and wood pegs. While these folks built they lived in huts made of sailcloth, bark, overturned boats, thatch, or whatever else they could find. As author Charles McRaven puts it, "History is full of accounts of civilized people thrust into the wilds, clutching remnants of their ordered, familiar pasts."[7]

And clutch they did. They constructed frames of thick beams perched on widely spaced, sturdy posts. A common dimension for the length and/or width of a building was 16 feet: a length based not on meticulous engineering calculations but on certain realities. A 16-foot log was about the most a man could horse around by himself. Spans longer than 16 feet started feeling bouncy, especially when used for loft floors. Some surmise that the 16-foot length, which commonly persists even today, is based on the width of four oxen, which might be lodged in a log barn, or the length of five axe handles, the measuring tape of the day.

The spaces between the posts were filled with rows of sticks and branches that were then coated with mud or clay, leaving the beams half exposed. This wattle-and-daub infill technique inspired the concepts of interior lath and plaster, and exterior stucco. When you see multimillion-dollar homes built today with a half-timbered look on the outside, be assured that the residents inside aren't putting up with snow, rain, dust, and bugs blowing through the cracks as their predecessors did. Eventually refinements were made. Clapboards—what we today would call lap siding—were applied to the outside, rustic wood paneling to the inside, and in between clay, straw, and bricks were added to block the elements. Eventually, vertical studs were installed between the timbers, followed by sheathing boards, then siding, to tighten up the structure even further.[8] Though not as durable as the brick and stone homes of Europe, about eighty seventeenth-century timber frame buildings are still standing in the United States today, the oldest being the Fairbanks House in Dedham, Massachusetts, built in 1636.

Three almost simultaneous inventions changed the face of housing forever. The first of those inventions was smaller than your little finger. Until around 1800, each and every nail was forged by a labor-intensive, hand-wrought process. After 1800, machines took over this task and produced nails faster and cheaper. About the same time, some sawmills converted from water power to steam power. Now sawyers didn't need to locate their mills on the banks of rivers and be dependent on the seasonal ebbs and flows of river currents. They could set up mills virtually anywhere. Standard-size lumber, as we know it today, became more readily available.

Third, a building concept known as balloon framing developed around 1830. Instead of using heavy timbers and tricky joinery for the framework of the house, balloon framing combined the first two inventions—nails and dimensional lumber—to create an entirely new building system. The lighter-weight easier-to-handle lumber, none thicker than 2 inches, could be used to build the floor, wall, and roof structures. Wall studs were so light that 16- or 20-foot-long studs could be used to create a continuous wall from sill plate to roof line, with the second story hung from these studs midway. These frameworks could be covered with 1-inch-thick sheathing boards to make them more rigid and at the same time block the elements and provide a nailing base for yet another layer of materials, like shingles or siding. Since the lumber was thinner, it could easily be joined with the now affordable nails. And this could all be accomplished by fewer people with fewer skills.

Legend has it that the term "balloon framing" evolved through the skepticism of some carpenters—a notoriously conservative lot when it comes to change—who thought this new-fangled form of construction was so lightweight and flimsy that structures would literally blow away like a balloon. But they didn't, and by the 1870s, most structures were built by use of balloon framing techniques.[9]

In the 1940s, yet another refinement, called platform framing, came along. This system involved building the first floor platform, then erecting the 8- or 9-foot-tall first floor studs, followed by another platform and then the second story walls, and so on. It allowed houses to be built even faster, using shorter, less expensive lumber and even less skilled carpenters. The floors separating each story provided not just a convenient work platform but built-in fire blocking to slow the spread of any potential fires from floor to floor. Plywood, pneumatic nail guns, asphalt shingles, standard-size windows and doors, and mass production made the process faster and easier yet.

During this long journey, styles ranged from rustic to Victorian to Craftsman to Bauhaus. But wood provided the skeleton.

In most books and classroom discussions you'll find the term "architecture" referring to buildings that (1) are square, (2) have floors made of something other than dirt, and (3) were built by someone of European descent. If you were a Plains Indian constructing a teepee out of poles and buffalo skins, you were pretty much left out of the history of architecture books. But anything lived in is architecture.

Early, early residential architecture moved in three directions—stone, earthen (brick), and wood—though frequently combinations of the three were used. In stone-rich wood-poor areas, dwellings were often made by stacking stone upon stone, with or without mortar. Bricks, both sun-dried and kiln-fired, were used in parts of the world where people had ready access to clay.

But where there was wood, it was used with abandon. In tropical areas, wood floors were slightly elevated on stilts to keep residents dry and safer from creepy-crawly things. Walls and roofs could be constructed in such a way as to keep out the sun, yet allow breezes in. In Western cultures, buildings may have been more substantial, but they were still constructed with the same goal in mind: to keep people dry, comfortable, and safe from creepy-crawly things.

Many early cultures revered trees. Wood was not a spiritless commodity. The Wanika of eastern Africa believed that trees, especially coconut trees, had a spirit and that cutting one down was equivalent to matricide. The Druids literally worshipped oak trees. In Germany, ancient laws dictated that any man who "dared to peel the bark of a standing tree" was punished in this manner: "The culprit's navel was to be cut out and nailed to that part of the tree which he had peeled, and he was to be driven round and round the tree till all his guts were wound about its trunk."[10]

Many primitive cultures believed that trees and the rest of the natural world were animate, and respected them accordingly. The ancient vegetarian Porphyry observed, "primitive man led an unhappy life, for their superstition did not stop at animals, but extended even to plants. For why should the slaughter of an ox or sheep be a greater wrong than the felling of a fir or an oak, seeing that a soul is implanted in these trees also."[11]

Most Americans have pretty much said "poppycock" to the whole soul-of-a-tree thing and take a more practical view. In the *Five Ages of Wood*, Michael O'Brien looked at the past four hundred years of North American housing history and broke them into five stages.[12] Most ages overlap, and all survive to some extent today. The basic notion is that we keep breaking wood down into smaller and smaller elements—from log to timber to two-by-four to ply to flake to fiber to cell—until we begin flirting with nanotechnology.

The first age of wood he describes as shaping, in which whole logs, used mostly in the form of log cabins, were the dominant structure. The second age—the age of joining—was represented by more permanently built timber frame structures. The third—the age of commodity—is symbolized by the two-by-four and other lumberyard boards used for stick frame houses. The fourth—the age of transformation—involves wood in severely altered states, such as oriented strand board, plywood, and engineered lumber. The age of reconstitution—the fifth and final division—is upon us and involves breaking down wood into individual fibers and cells and then combining them with other chemicals and components to create composite materials such as Trex decking.

O'Brien fears that as we reduce wood into smaller and smaller components we also reduce our emotional attachment to the structures and our sense of place. He may be right. I still remember the gigantic owl-eye grain pattern on the birch closet doors of my childhood bedroom; they spooked me sometimes, but I still felt an attachment. I feel little attachment to oriented strand board today. The warning O'Brien issues is that future houses are in danger of becoming more like machines—isolated space ships, sitting porchless and maintenance-free on suburban lots—rather than things connected to, and with, their surroundings and occupants.

Even in a nanotech world, the log cabin persists. Today there are over half a million log homes in the United States, and thirty thousand are added annually. Structures range from $5,000 do-it-yourself kits to the $10-million-dollar log home of former Tyco CEO Dennis

Kozlowski (that, without the $6,000 shower curtain paid for by shareholders). And they're structures where your family can gather around a cozy hearth and play checkers—comfortably.

EVERYTHING YOU NEVER WANTED TO KNOW ABOUT CONSTRUCTION LUMBER

We can avoid a lot of confusion right off the bat by understanding that most terms in the world of lumber and construction are mystifying acronyms, grossly misleading or outright wrong. A 2 × 8 × 12 isn't two or eight or twelve of anything. A 16d or 16-penny nail never has, and never will, cost 16 pennies. Half-inch plywood is rarely 1/2 inch thick, and though the lumber you buy may be stamped hem-fir, there's not a hem-fir tree on the planet. The 2 × 12 you pick out of the lumber pile on Monday may be able, according to building codes, to span 2 feet farther than the 2 × 12 you pick out of the lumber pile on Friday.

Yet, given all of this uncertainty and chaos, there are some basic principles one can apply to wood and wood frame structures. The first three principles have to do with the weight lumber can support, based on its orientation. In laymen's terms, it boils down to this:

1. Wood is *really, really strong* in compression parallel to the grain, as when it's used as a stud in a wall or as a post holding up a deck. For example, a single high-grade 4 × 4 Douglas fir post can support nearly 10 tons.[13] Bruce Hoadley, author of *Understanding Wood* (and *the* man when it comes to wood science) states, "I can never remember ever seeing a structural failure or even hearing of one due purely to compression stress parallel to the grain. If compression were the only factor, a 250-pound person could be supported by four hickory dowels, each one only 1/8 inch."[14] When Hoadley talks about compression being the only factor, he's talking about preventing

the weakness and failures that occur when vertical posts or studs start bowing. The sheathing that covers studs prevents them from bowing, and single posts are often stabilized by angle-braces to prevent bowing.

2. Lumber is *pretty strong* when it's standing on edge, like a floor joist or a support beam. The deeper a piece of lumber, the farther it can safely span. In the case of joists, 2 × 6s placed every 16 inches can safely span about 9 feet, while 2 × 12s spaced the same distance apart can span over twice that distance.[15] The board's species and its grade (based in part on how many large knots the board has) will affect this span. Douglas fir or larch will span much farther than lodgepole pine or cedar. Another element of joists has to do with the modulus of elasticity, a measure of stiffness. The higher the number, the stiffer the wood, and the more solid a floor feels.
3. Lumber is *wimpy as hell* when it's "lying flat," like a deck top board or flat plank you might use as a ramp. The same 2 × 6 that can span 9 feet on edge can span only 24 to 30 inches, according to most building codes, when lying flat like a deck board.

When a conventional wood frame house is built, the strengths are capitalized upon and the weaknesses are accommodated. Since wood is so strong in compression parallel to the grain, wall studs need to be only 2 × 4s to hold a house up; 2 × 6s might be used to increase the room available for insulation or plumbing pipes, but they're rarely needed for strength. Since wood is pretty strong standing on edge, spacing joists every 16 or 24 inches is optimum because that's close enough to support the plywood subfloor that will be attached without having it sag; just as important, it allows the joists to span from wall to wall or from wall to beam without the joists themselves sagging or overflexing. And since lumber is wimpy lying flat, today it normally isn't used that way in residential construction except for decking on an outside deck.

Other important elements are required for wood structures to remain strong. They need to include wind-bracing. You can think of wind as a sort of horizontal gravity that applies itself at sporadic intervals. All structures—big and small—need to account for this force. In tall, big city buildings, up to 10 percent of the structural weight may go into providing wind-bracing, but this would be overkill in wood frame houses.[16] To prevent racking, 1 × 4 diagonal cross-bracing is often "let in" or notched into the wall studs at each end; sometimes metal strapping is also used. The plywood and other types of sheathing on the walls, when installed and nailed properly, also act as wind-bracing, effectively creating zillions of smaller triangles all over the exterior wall surface.

Wood structures are also durable because they possess a property termed redundancy. In the book *Why Buildings Fall Down,* the authors explain, "Structural redundancy essentially allows the loads to be carried in more than one way—i.e., through more than one path through the structure—and must be considered a needed characteristic in any large structure or any structure whose failure may cause extensive damage or loss of life."[17] The book goes on to describe disasters that were or were not avoided based on redundancy: The Empire State Building exhibited great redundancy by barely budging when a B-25 bomber crashed into its seventy-ninth floor in 1945. Though one support column was damaged, the others, spaced at 19-foot intervals picked up the slack "like a centipede that can compensate for the loss of a leg by redistributing its weight to the remaining legs."[18] When the walkways at the Hyatt Regency in Kansas City came crashing down in 1981, killing 114 people, the problem boiled down to lack of redundancy in the connection between the hanger rods and walkway supports which gave way. When the connections gave way, there was no backup.

If you look closely at the wood frame structure of a house, it's easy to spot this redundancy. (Instead of "redundancy," some would simply use the term "overbuilt.") Studs, joists, and rafters are spaced every 16 to 24 inches; if some are damaged, overstressed, or notched (a favorite trick of plumbers), the others easily carry the extra load. The sheathing on the walls, ceilings, and floors also helps distribute the weight to a broader area. The horizontal top and bottom nailing plates hang everything together. Blocking between some of the members transfers weight to adjacent members. Wood structures may fall off their masonry foundations or rot through neglect or slowly sag and lean, but a catastrophic failure of the type witnessed at the Hyatt is virtually unheard of. Wood structures give you plenty of warning before they fall down.

Wood structures are also resilient—not resilient in the mushy sense of a tennis ball, but rather as a thing that has "give." When a stupendous earthquake shook Charleston, South Carolina, in 1886, fourteen thousand chimneys fell, and nearly every brick building in town was damaged. But one family wasn't even aware of the catastrophe until they arose and stepped out of their house—their wood house—the next morning. Component masonry—the type built brick by brick or block by block—is very strong in compression, just like wood, but lateral or sideways movement can be ruinous (though concrete reinforced and prestressed concrete structures fare extremely well). Wood structures can move to and fro, but component masonry moves well only "to"; on the "fro" it can collapse. Wood structures have "give," but that "give" is tied very well together with nails, sheathing, top plates, and even drywall.

So wood frame structures have lots of positive traits—strength, resiliency, and redundancy. But they have several Achilles' heels.

If you keep wood in a perfect environment, it will last forever. When moisture, insects, mold, fire, abrasives, and other elements are kept away from wood, or when coatings or preservatives are used to protect it, wood can last thousands of years. However, when you put all these limitations on it, wood becomes a museum piece, not a hardworking member of society. Wood structures may be resilient in earthquakes, but when natural gas lines break and spew fire, and broken water mains can't quench the flames, they burn. Wood is an organic substance, which is a splendid way for Mother Nature to hand us building materials. But things that grow in nature eventually get consumed by nature.

A final characteristic of wood and wood structures involves their energy efficiency—in terms of both the energy required to produce the raw product and the energy required to heat the finished product. On a ton-by-ton basis, compared with the energy required to convert a tree into lumber, it requires 5 times as much energy to produce a ton of cement, 14 times as much to create a ton of glass, 24 times as much for a ton of steel, and a whopping 126 times more energy for a ton of aluminum. Though wood products make up 47 percent of all the raw materials used in manufacturing in the United States, that process represents only 4 percent of the nation's manufacturing energy bill.[19]

Wood also has relatively good insulating qualities. R-value is a numerical value given to building materials to indicate their ability to reduce the movement of heat through them. While plywood and lumber have an R-value of around 1.25 per inch, brick is a sniveling 0.20, concrete a bone-chilling 0.08, and stucco a paltry 0.02.

Wood, despite its imperfections, remains the perfect material for many uses.

A DIRTY ROTTING SHAME

When workers began investigating moisture problems with one house in Woodbury, Minnesota, they found rotted-out plywood with the thickness and structural integrity of paper, 2 × 4 studs that looked like waterlogged driftwood, and rotted windowsills that could be penetrated 3 inches with an awl. You might expect this from a neglected eighty-year-old fixer-upper, but this house was far from that. This house was a youthful seven years of age. One inspection company gave the owners a repair estimate of $400,000 on a house that originally cost $550,000 to build.[20] The problem didn't originate with the plywood or the 2 × 4s but rather on the outside, with the synthetic stucco—a product generally referred to as exterior insulating finishing system, or EIFS. More correctly, the problem was with the way the EIFS covering was installed. Any water infiltrating around

Wall sheathing on a six-year-old house that was clad with exterior insulating finishing system, showing serious rot.

poorly flashed windows or seams could get in but couldn't get out.

Rot needs four elements to thrive: mold spores (which are everywhere), something to eat (wood), a temperature range between 40 and 104 degrees F (I know I keep my house in that temperature range), and moisture. Long-lasting wood structures last long because they don't let moisture in, and the moisture that does get in can easily escape. This suburban house didn't let it escape.

How serious is the problem? Thousands of lawsuits have been filed by dismayed new homeowners, leading one inspector to flatly state, "All [synthetic stucco] houses are simply doomed. I've never seen one that will outlive a twenty-year mortgage."[21] Mac Pearce, a microbiologist specializing in mold in homes, bemoans the situation, saying, "We're building self-composting houses."[22] While these in all likelihood are overstatements, the bottom line is this: When wet wood can't breathe properly, it rots.

* * *

We old house dwellers may cry like workers in an onion-chopping plant when we open our monthly utility bills, but I'll tell you one thing—our houses breathe. In fact, our 155-year-old house hyperventilates. Old houses may go through fifty to seventy complete air exchanges in a day, while a new tightly sealed home may have just one. Both numbers are far from perfect. Fifty air exchanges per day means you're heating fifty houses' worth of air each day in order to keep one house warm. In an old leaky house, you don't have to admonish your kids to "close the door; you're heating the whole outside." You're doing it anyway.

But only one air exchange can be worse. It means that all the moisture generated inside from a family of four on any given day—from showers (1/2 pound), cooking and dishwashing (3 pounds), washing clothes (1/2 pound), mopping the floor (3 pounds), watering plants (3 to 5 pounds), or simply living and breathing (12 pounds)—stays inside. It throws itself against windows, tries to crawl out through little gaps around your outlets, and condenses inside your walls. Add 2 to 5 gallons of water per day generated by building components drying out when a house is first built, along with the possibility of hundreds of gallons coming in through faulty flashings or damp concrete basement floors, and you've got a rotten problem to solve.[23]

Plastic vapor barriers, high-tech caulks, windows and doors that fit tight as a cork, synthetic stuccos, sealed furnaces, foam sheathing, and other new building materials trap the heat inside, but they also trap moisture inside the house or, worse yet, inside the walls. One company has gone so far as to train Mold Dogs™ to sniff out the problems. In the cozy words of mycologist Dr. George Carroll, "The appearance of mould-sniffing dogs is a brilliant solution to the problem of locating sources of household mould. Dogs when properly trained are in fact exquisitely sensitive walking gas chromatographs with a proven record of locating mouldy substrates even when hidden behind partitions or on the insides of walls."[24]

While discovering moisture problems early on is good, preventing them is better. Building scientists recommend somewhere in the neighborhood of eight air exchanges per day as the optimum balance between conserving energy and venting moisture. You can crack a window, turn on a bath fan, or, if you're building a new house, ask your builder to "build 'em like they did in the 1850s." But that doesn't sit well with most building codes or inspectors. There are better solutions. Air-to-air heat exchangers and heat recovery ventilation systems pull fresh air in from the outside and use the heat from the exhausted air to help warm the incoming. And in a day and age when most people spend over 90 percent of their time indoors, and that indoor air can be five times as polluted as the air outside, the clean fresh air you get as a bonus will help *you* stay healthy, too.

But an EIFS isn't the only thing that can lead to rotting wood and rotting property values. Any house that's built with materials or methods that won't let moisture escape is going to have problems. Water can infiltrate via bad roof flashing, poor drainage around the foundation, and poor connections where decks meet the house. Dryers that vent moisture directly into the house or attic space (I've seen that) can dump 25 gallons of water into the house on washday.

Bill Rose, of the Building Research Council at the University of Illinois, thinks it boils down to this: "Basically we're trying to build with unskilled labor, and we've probably lost the connection to the old guys who knew how to keep water out of a building."[25] The "old guys" who built our old house are smiling somewhere.

WINCHESTER HOUSE: THE THIRTY-SIX-YEAR REMODELING PROJECT

If you think your neighbor's kitchen remodeling project has gone on too long, be thankful you never lived next door to Sarah Winchester. *There* was a woman who liked renovation. *There* was a woman who would have been glued to *Extreme Makeover, Home Edition.*

Sarah Winchester became heir to the Winchester Repeating Arms Company fortune in 1881 after her husband, William, died of tuberculosis. This tragedy, in conjunction with the death of their infant daughter fifteen years earlier, left Sarah reclusive, alone, and rich—to

The "Door to Nowhere"—one of the many, many quirky architectual elements of the Winchester House meant to confuse and baffle vengeful spirits of those killed by Winchester rifles.

the tune of $20 million, plus 49 percent ownership of the gun manufacturing company to boot. Shortly after her husband's death, Sarah visited a medium, who told her there was a curse on her family that had taken the lives of both her husband and her daughter. "It will soon take you too. It is a curse that has resulted from the terrible weapon created by the Winchester family," the medium (most likely not a card-carrying member of the NRA) intoned. "Thousands of persons have died because of it, and their spirits are now seeking vengeance."[26]

Sarah was told to sell her East Coast home and head west to start a new life. In fact, she was told to not only start a new life but start a new building project as well. The medium told her to "build a home for yourself and for the spirits who have fallen from this terrible weapon, too. You can never stop building the house. If you continue building, you will live. Stop and you will die."

Apparently Sarah Winchester was listening closely. In 1884, she purchased 162 acres and a house already under construction in Santa Clara County, California. She then proceeded to build nonstop, twenty-four hours a day, for the next thirty-six years. She had no master plan but rather met with the foreman each morning to discuss her hand-sketched plans for the day. At its pinnacle, the house reached seven stories in height and had been expanded to 160 rooms, including 40 bedrooms, 2 ballrooms, and 47 fireplaces.

In that era, nearly every component of a house was made of wood: pine framework and sheathing; redwood siding and shutters; cedar shingles; hardwood flooring, trim, paneling, and doors. So much lumber was required that a railroad spur was diverted to the house.

Since the vengeful spirits killed by Winchester rifles were out to wreak havoc on her life, Sarah's designs were based on attempts to wreak havoc on theirs. She included countless stairways and hallways that led to nowhere, doors that opened to blank walls, and trapdoors. All the stair posts were installed upside down; the bathrooms had glass doors; and my oh my, talk about an obsession with the number thirteen. The windows had thirteen panes, clothes hooks were in multiples of thirteen, the greenhouse had thirteen cupolas, and all the staircases but one had thirteen steps. The "normal" staircase had forty-two steps, but each was only 2 inches high. To honor this obsession, the groundskeepers today have created a topiary tree shaped like the number thirteen.

The San Francisco earthquake of 1906 greatly shook both Sarah and the house. Though the house fared better than most, the top three floors of the house collapsed. She was convinced the earthquake was a sign from the spirits indicating their displeasure that construction seemed to be drawing to a close. In response, she boarded up thirty rooms so they could never be completed (which also served to forever sequester the spirits trapped in those rooms.) Then she proceeded to keep building until her death in 1922.

When Sarah's heir, her niece Frances Marriot, went to claim her fortune, she discovered a sum far less than the $20 million Sarah had started out with. In fact, she discovered not much at all, above and beyond the house itself. The rumors of vast amounts of jewelry and solid-gold dinner service hidden in the house turned out to be just that—rumors. When the numerous safes throughout the house were forced open, only fish line, socks, a few newspaper clippings, and a suit of woolen underwear were found. The movers in charge of emptying the house became so lost in the labyrinth of rooms that it took them six weeks simply to get the furniture out, and every time a room count was done over the next five years, there was a different total.

The house has been declared a California Historical Landmark and is open to the public for a rather spooky admission price of $26.95. You can even take one of the flashlight tours offered at Halloween and every Friday the 13th—but not if you're afraid of ghosts. Employees and visitors still witness banging doors, mysterious footsteps, and doorknobs that turn by themselves.

* * *

Wood has been an indispensable ally in keeping us warm, dry, and secure. But not only has it sheltered us, also it has provided us with many of the things we use on a daily basis—things both great and small. Whether it's toothpicks or wine barrels, pencils or kidnappers' ladders, wood has stuck with us the whole way: ways like those shown in the next chapter.

CHAPTER 7

Wood in Day-to-Day Life

In his classic children's book *The Giving Tree*, Shel Silverstein takes a look at the relationship between a boy and a tree. At first, the boy asks only for shade and permission to swing from its branches. And "Oh the tree was happy. Oh, the tree was glad." As the boy grows he makes increasingly taxing requests of the tree—and the tree always graciously accommodates him. When the boy needs money, the tree gives him apples to sell. When the boy grows older and needs a house, the tree gives him branches to build with. When the boy becomes depressed, his life "turned so cold," the tree gives up its trunk so the boy can build a boat and sail away. And when the boy finally returns as an old man, and the tree is nothing but a stump, the tree still gives. "Come, boy," the tree says, "Sit down, sit down and rest a while." An old stump is still good for that.[1]

WHEN WOOD WAS EVERYTHING AND EVERYTHING WAS WOOD

A heart-warming tale, to be sure. But *The Giving Tree* lives in a lot of places besides the children's fiction section. We're all that little boy. We ask, and the tree provides. "Oh the tree was happy. Oh, the tree was glad."

Until a few generations back, wood was the raw material used to build just about everything. In *A Natural History of Trees*, Donald Peattie writes:

> The century of magnificent awareness preceding the Civil War was the age of wood. Wood was not accepted simply as the material for building a new nation—it was an inspiration. Gentle to the touch, exquisite to contemplate, tractable in creative hands, stronger by weight than iron, wood was as William Penn had said "a substance with soul." It spanned rivers for man; it built his home and heated it in the winter; man walked on wood, slept in it, sat on wooden chairs at wooden tables, drank and ate the fruits of trees from wooden cups and dishes. From cradle of wood to coffin of wood, the life of man was encircled by it.[2]

Poet W. H. Auden honored the material in penning "A culture is no better than its woods."[3] *The International Book of Wood* flatly states, "Man has no older or deeper debt than that which he owes to trees and their wood."[4]

Here are six more "once upon a time" stories wherein wood has served as the Giving Tree.

ONCE UPON A TIME WOOD WAS A KICK-ASS CONFIDENCE BUILDER. Wood in humankind's early development simply made sense. The combination of happenstance and primitive logic made the use of wood a natural. It met people's basic needs. Though surviving wooden artifacts are largely nonexistent, one could see its use evolving in storybook fashion like this: Guy Uno is walking along the river and sees fruit on yonder side. The river is swift, but that fruit looks good. He takes a step in, gets knocked off his already unsteady feet, and begins foundering. A log zips by, and he grabs it. Suddenly he's floating. He's floating away from the fruit, but at least he's not floating away dead. There's his boat.

He gets out and dries himself along the edge of a smoldering clump of trees struck by lightning. Hmmm—maybe someday there will be a use for *that.* He begins walking. Again he's tempted by the fruit, but he's not risking a second dunking. Then he happens upon another tree—this, a tree that's fallen from one bank to the other. And he sees a furry thing scurry across it. All of a sudden he sees a safer path to his fruit. He hasn't invented the bridge, but he has discovered it. And for future crossings he may be tempted to drag a fallen log over to the riverbank, toss one end to the far side, and make his own bridge.

Eventually, humankind would discover that stone laid upon stone made a sturdier bridge, and that sand mixed with limestone made a longer-lasting bridge, and that the shiny substance gleaned from rock, when refined a certain way, made a longer spanning bridge. But none of these raw materials was as accessible and adaptable as wood.

Wood—whether in the guise of a log-inspired boat, a trunk-inspired bridge, a branch-inspired shelter, a stick-inspired spear, a forest fire–inspired campfire, a twig-inspired toothbrush, birch bark–inspired paper, or a log-inspired wheel—lent itself to easy adaptation and discovery. It led to an easier next step. It was, in short, a kick-ass confidence builder.

Wood was critical in man's expansion of his geographic and mental boundaries; it gave him the energy and substance to flourish. Wood helped propel man out of his loincloth into a three-piece suit.

ONCE UPON A TIME WOOD WAS THE PRIMARY FUEL. Vast amounts of wood have been consumed throughout history for building houses, ships, and tools, but inconceivable amounts have been consumed as fuel: initially in the form of firewood, later as charcoal. During the heyday of copper smelting in Cyprus in the fourteenth century BC, 4 acres of pine trees were needed to produce the 6 tons of charcoal required to refine just one ingot of copper. In sixteenth-century England, blast furnaces used to smelt and refine iron consumed forty-eight cords of wood—the equivalent of 140 tons of freshly cut oak—to create a single ton of iron.[5]

Wood as a fuel was so critical that some have linked the decline of the Roman Empire to the decline of its forests. Pottery, metal smelting, glassmaking—even the hallowed Roman baths—all had gluttonous appetites for wood. It's estimated that silver-smelting furnaces of the Empire alone consumed over half a billion trees during four centuries of operation. Forests were rapidly depleted.[6] By the third century, Roman wood merchants were traveling as far as northern Africa in search of wood.

ONCE UPON A TIME WOOD WAS AS PRECIOUS AS GOLD. Wood was a precious commodity. Civilizations with access to bountiful forests were able to build exceptional naval fleets for exploring, craft superior weapons for conquering, and create vast amounts of charcoal for manufacturing. Taking full advantage of the woods of newly discovered lands was the norm, sometimes even the motivation. Woods were sequestered in royal treasuries, battles were fought over strategic wood reserves, and wood trade routes were vigorously defended.

Once Europeans discovered Central America's forests of teak—a valuable wood for shipbuilding—they realized there were other types of gold in the New World. Half a world away in New Zealand, over one billion board feet of kauri wood were harvested within the first twenty years of European settlement. In pioneer America, the seemingly endless supply of wood spurred the development of railroads, telegraph lines, and charcoal-fired mills. By the later part of the nineteenth century, lumber production rose to 35 billion board feet annually.[7]

ONCE UPON A TIME (EVEN *RECENTLY* UPON A TIME) WOOD HELPED FORGE NEW FRONTIERS. In a new world, a person could head into the forest and make a go of it with little more than an axe slung over his shoulder and wood-headed determination.

If you were smart enough to settle in a mixed hardwood forest near a stand of conifers, your task was made all that much easier. From the pines you could craft a log cabin. From the cedar you could rive shingles for roofing and boards for a door. From the hickory you could craft solid hinges and locks. From that same hickory you could build chairs and a bed, and weave the bark into chair seats and rustic bedsprings. You could chop birch for firewood and use the bark for kindling, all burning warmly in a clay-lined log chimney. You could fashion a maul of hickory, then use it, along with locust wedges, for cleaving cedar logs into rails to create split rail fencing for domesticated animals.

You could split basswood into thin sheets, score it, and bend it into boxes. Wagon axles and wheels could be gleaned from hickory. If you needed dye, you could make deep brown from butternut and dark yellow from chestnut oak. From white oak you could make watertight barrels, tankards, and churns. With the bark of that oak you could extract tannins for making hides softer and more durable. From willow you could weave baskets for storage, and from the ironlike persimmon you could make shuttles for weaving. From maple you could make an indestructible chopping block for butchering and chopping.

You could use sawdust to make ice blocks last well into summer, and fireplace ashes to turn animal fat into soap. Fruits could be dried and nuts stored to stock a hearty pantry. And when the day's work was done, you could smoke a fine cherry pipe while whittling a butter scoop out of apple wood. All of this you gleaned from the forest with a simple axe.

The importance of wood in carving new frontiers isn't just a thing of the distant past. In 1915, Joseph Knowles, in cahoots with the *Boston Post*—at that time in a circulation war with Randolph Hearst's *American*—headed into the woods of Maine for a sixty-day stint (some would call it stunt), wearing only a jockstrap, in his version of a twentieth-century frontiersman. He carried nothing. He emerged two months later looking like Tarzan. In that two months it was wood that sustained him. He built a crude shelter out of sticks and covered it with pine boughs and bark. He made fire using stick friction and bark tinder. He clobbered the frogs he ate with wooden clubs and caught partridges with a slip noose made of bark. He crafted cigarettes out of squaw bark and leaves, made moccasins out of bark, and crafted a bow and arrows from hornbeam wood.

As fall approached and the need for a coat became apparent, he dug a deadfall pit, using sticks, then covered it with more sticks. He used a wood club to kill the young bear that fell into the trap. After which, he skinned the bear and made its furry skin into a coat for himself. And he recorded the entire affair on birch bark, using the charred end of a burned stick.

People—even in 1915—thought it was cool. Five days after striding out of the woods, Knowles was cheered by twenty thousand Bostonians as he paraded down Newspaper Row. That was after he paid the $205 fine to the bureaucrats at the Fish and Game Commission for hunting out of season and "making fire without a license."[8]

ONCE UPON A TIME THERE WAS A GUY WHO LIKED WOODEN THINGS SO MUCH THAT HE SAVED THEM FOR ALL OF US TO SEE. If you wish to step back into the era of wood, step through the doors of the Mercer Museum in Doylestown, Pennsylvania. It houses collections that give you appreciation for the vital role played by obsessive-compulsive people in the perpetuation of the species—or at least of its artifacts. Within the six-story concrete castle Henry Mercer built to preserve and display the artifacts of everyday American life of the eighteenth and nineteenth centuries, you'll find fifty-five exhibit rooms—each dedicated to a specific trade or craft—packed with forty thousand objects. And here, with history distilled down to its basic elements, you can see the overwhelming role of wood.

There are wooden spinning wheels, shoe lasts, sleighs, wagons, whale boats, woodblocks for printing wallpaper, cigar molds, baskets, prosthetic limbs, weaving shuttles, plows, sawmills, cigar-store Indians, butter churns, Conestoga wagons, barrels, grain shovels, and hurdy-gurdies.

Though Mercer strove to collect the everyday (the collection includes sixty apple corers), the unusual sneaked in. There is a calithumpion—a raucous noise-making device that was traditionally banged outside newlyweds' windows on their wedding night. There is a terrapin scoop used to probe for, then scoop up, fresh water turtles, considered a delicacy. There's a hog yoke; a large wooden

The Mercer Museum—a history museum of everyday life in the eighteenth and nineteenth centuries—displays tens of thousands of wood implements and artifacts from that era. Whether in whaleboats, wagons, or sleds, wood was in the thick of things.

wishbone that was secured around a hog's neck, making it difficult for the hog to sneak through a fence. The display gives visitors a tangible sense of their ancestral woodiness.[9]

WOOD IS STILL CRITICAL—AND THAT'S NO FAIRY TALE. On a worldwide basis, wood consumption remains on a tear. An average of 3 1/2 pounds of wood is consumed each day by each of the six billion people walking the globe. To sustain this pace, nearly 4 billion tons of wood needs to be harvested from the earth's forests annually.[10]

This wood is consumed in different forms, by different groups, in different parts of the globe. About 55 percent of the world's wood is used for cooking and heating, primarily by over two billion people in developing countries. About 30 percent is used to manufacture paper products—everything from cardboard boxes to toilet paper—and is consumed primarily by those in industrialized and developed countries. It's only the remaining 15 percent—the percent most of us think of as "wood"— that's used for housing, furniture, and other solid products.[11]

I ran across a list of wood products titled "Wood You Believe," ever so cleverly formatted in the shape of the classic pyramidal Christmas tree.[12] While many of the 171 items were obvious—toilet seats, guitars, church pews, stepladders—many others weren't. It listed things like these:

Ping-pong balls and rayon. These cellulose-based products are made from raw material derived from wood (or cotton).

Linoleum. The rosin used as a binding agent comes from pine trees, and the wood flour used to ensure colorfastness and a smooth surface comes from hardwood trees.

Safety glass. The earliest versions consisted of two layers of glass laminated together with a thin layer of tree resin.

Imitation bacon. Torula yeast, which grows on wood sugars left over from the pulp-making process, is a key ingredient in foods ranging from baby foods to cereal to imitation bacon.

Cosmetics. Oh, let us count the ways. Coconut butter as a skin softener; camphor as a coolant and stimulant; eucalyptus oil as an antiseptic, *Fagus silvatica* from the beech tree to help remove wrinkles, wintergreen oil from the *Betula lenta* as a pain reliever. When you start adding in derivatives from nuts, seeds, and flowers, the list careens into the hundreds.

Cancer cures. Taxol, a derivative of Pacific yew bark, is a medical application that shows great promise, but it takes a lot of bark. One study shows that it takes 30 pounds of bark—about one and a half tree's worth—to create 1 gram of taxol. A typical series of four treatments requires a total of 2 grams.[13] Based on its use to treat ovarian and breast cancer, a million and a half trees would need to be harvested annually. Four million yews exist today on federal land.

Many sources quote five thousand as the number of products derived from trees today, but when you throw in the fish floats, football helmets, and fireworks derived from wood products, one might call this a conservative estimate. It's a material with options. If you owned a cord of wood, you could manufacture twelve dining room tables, 4,384,000 postage stamps, one twentieth of a house, or three hundred copies of the book you're now reading.[14]

Robert Youngs summed up the ever-vital role of wood: "If we are running toward the goal of supplying the greatest good for the greatest number of people for the longest time, wood as a resource for materials and energy matches all running mates . . . the age of wood is both a time-to-come and a past era. The Age of Wood was, is and will be Every Age."[15]

THE LINDBERGH KIDNAPPING, THE TED BUNDY TREE, AND FORENSIC WOOD

No fingerprints, no footprints, no witnesses. The only solid piece of evidence left at the scene of the kidnapping was a homemade wood extension ladder. But that was all Arthur Koehler needed to link the suspect to the crime. Slivers and tool marks were analyzed, nail holes were measured and matched, and in the end the bad guy was convicted. A scene out of a 2007 episode of *CSI*? No, a scene out of the 1932 kidnapping case of Charles Lindbergh's two-year-old son, considered by many to be the crime of the century.

In 1932, Charles Lindbergh was the toast of the town. His solo 1927 transatlantic flight had made him a national hero. He had it all: Fame, fortune, a Congressional Medal of Honor, a New Jersey estate, a beautiful wife, and a healthy two-year-old son, Charles Lindbergh, Jr. This idyllic life all changed the evening of March 1, when the family's nanny went into the nursery to check on Charles Jr. at 10 p.m. only to find no one to check on.

The part of the story relating to the botched police investigation, the series of ransom notes, and the eventual discovery of Charles Jr.'s badly decomposed body is one of high intrigue and speculation. But it's the three wood ladder sections, each 80 inches long, with rungs spaced 19 inches apart, that interest us here.

Investigators had followed dead-end clue after dead-end clue, and the Lindberghs and the public at large were growing impatient for a conviction. Arthur Koehler, chief wood technologist at the Forest Products Laboratory (FPL), felt that given the chance, he could make the ladder—"the wood witness"—talk. In late May 1932, he was given the chance. He analyzed slivers of wood sent to his office and determined that the ladder had been built from four types of wood: North Carolina pine, West Coast Douglas fir, birch, and ponderosa pine. Interesting data, but data that seemed to point nowhere. Ten months later, Koehler was asked by the head of the New Jersey State Police, a man by the name of Colonel H. Norman Schwarzkopf—yes, father to Stormin' Norman—to conduct a more thorough examination of the ladder. And thoroughness is what they got.[16]

Koehler noted during his investigation that the ladder had the strange qualities of being ingeniously designed but crudely constructed.[17] After a few days of preliminary investigation in New Jersey, Koehler headed back to his laboratory in Madison, Wisconsin—ladder in the aisle next to him on the train. He measured each of the nineteen ladder parts to the nearest 1,000 inch. He angle-lit each board to highlight marks left by saws, planes and chisels, and recorded the information. He found virtually no signs of wear, indicating that the ladder had been built for this one single mission. By matching grain patterns, he was able to determine that ten of the rungs had been cut from a single 1-inch by 6-inch ponderosa pine board. Two of the long ladder sides, or rails, had come from one 1-inch by 4-inch by 14-foot-long North Carolina pine board, while another rail—which was referred to as the now-infamous "Rail 16"— had been trimmed down from a wider board of some sort.

The Lindbergh kidnapping was the first case to take advantage of forensic botany— the umbrella name given to the field of scientific research in which plant science is used to resolve legal questions. Thousands of cases have followed suit, and the Forest Products Laboratory has been involved in many of them.

Seventy-three years after Kohler began his investigation, I'm standing in front of a gigantic poster titled "Wood Anatomy in Forensic Science" mounted to a wall in the Forest Products Laboratory—the same laboratory where Koehler was employed. Next to me, wood anatomist Reggie Miller explains the three cases portrayed in pictorial form on the poster. He was involved in all three.

The first shows a sawed-off shotgun found in the trunk of a car. The suspect steadfastly denied that the gun was his, since he, as a convicted felon, would be sent to jail for life if it was proved the gun was his. The point-blank evidence was the sawed-off butt of the rifle found in the guy's wastebasket—with growth rings that matched exactly. "Any idiot could see that these pieces match up, but [the court] needed an expert to tell them that," explains Miller. The guy was convicted.

The second series of photos have to do with the Ted Bundy Tree. Bundy, suspected of killing as many as forty young women in the 1970s, was being readied for execution in Florida in 1989 when pulpwood cutters in a national forest in Utah came across a tree that had "Ted Bundy" carved into it, along with a date that looked like "'78." A girl had disappeared in that very area in 1978, and since it was known that Bundy had been in the area at the time, he became a prime suspect. Bundy wouldn't confess to the crime, and officials couldn't locate the body, but a degree of doubt could be eliminated by determining when the initials had been carved. Inasmuch as Bundy had been in custody since mid-February 1978, if it could be proved that the initials were carved after that time, the initials would be proved to be a forgery and hoax. Slices of wood from the initialed area were sent to the Forest Products Laboratory and the Tree Ring Lab in Arizona. Miller explained that scientists looked at the bark and the wood tissue being laid down in annual increments and determined the initials had been carved prior to Bundy's capture. They never found the body, and Bundy remains a suspect to this day.

The third series of photos documents how the FPL helped crack a murder case on a Georgia military base. Fragments of wood with paint chips were found on a woman's body, and it was thought that if the wood fragments could be matched to that of the wood inside the suspect's house (the suspect was the husband), that would constitute evidence as to where the murder was committed, and therefore who committed it. The FPL was able to match not only the wood on the body with wood found in the basement, but also the pattern of the fungi growing on the woods. Based on this and other evidence, the guy confessed.

Koehler had a much rougher time of things in the Lindbergh case. In a game plan that would seem far-fetched even in today's high-tech world, he surmised that if he could figure out the exact characteristics of the planer used to mill the 1-inch by 4-inch by 14-foot board used for the rails, he'd have at least a toehold on the Lindbergh kidnapper. He found that the planer had six knives in the side heads—one that left a distinctive mark—and eight knives in the heads that planed the top and bottom surfaces. The heads rotated at 3,300 revolutions per minute, and the automatic feed mechanism pushed the board through at a rate of 258—not 255 or 260, but 258—feet per minute. He accumulated data about the growth rings and exact dimensions of the board, then tabulated all this information into a three-page letter and shipped off inquiries to the 1,598 mills in the region that handled North Carolina pine.

The process of elimination began. You only mill 2-inch material? You're out. Your planer has four, not six, knives in the side heads? Take a step back. Your feed mechanism pushes boards through at 220 feet per minute? Not you. Eventually the field was narrowed to 25 mills, and each submitted a one-by-four to Koehler for examination.

One board had distinct markings in the same area where the plane that milled two of the ladder rails had left distinctive marks—a clue that perhaps that cutter blade had been recently replaced. He visited the mill of the Dorn Lumber Company and, based on information gleaned there, not only determined that the board had been milled at their facility, but also deduced the most probable lumberyards to which they had shipped the boards.

After a half year of "frustration, dead ends [and] stares of disbelief,"[18] Koehler was finally able to narrow the retail outlet down to the National Lumber and Millwork Company in Bronx, New York. Eureka! If investigators could go through the company's books, they could check out customers who had purchased one-by-fours within the given time frame, and perhaps close in on the kidnappers. But Eureka! The store operated only on a cash-and-carry basis. It kept precious few records. The field, at best, had been narrowed only to the 1.3 million people living in the Bronx.

But this high-profile story continued to grip the nation even after eighteen months. Then, in September 1934, a carpenter of German heritage, Bruno Hauptmann, was arrested for passing some of the marked ransom money. The $14,000 in marked ransom money found in the garage was surely incriminating, but even more incriminating was something else in the garage: Hauptmann's tools. The nicks and grooves in the blade of his 2 1/2-inch hand plane created the exact same nicks and grooves as those found on the ladder rails. His handsaw blade cut the exact same 0.0035-inch-wide kerf as the kerfs cut on parts of the ladder. Nails with the letter P imprinted on the shaft were the exact same type used to build the ladder. A pencil sketch of the ladder used in the kidnapping was found in one of Hauptmann's notebooks.

But an attic floorboard, or rather part of a floorboard, sealed Hauptmann's fate. It was determined that Rail 16 of the ladder and a floorboard found in Hauptmann's attic were originally part of the same board. Hauptmann, out of either stinginess or wont, had sawn off and removed part of a floorboard in the attic to create one of the ladder rails. When Rail 16 and the floorboard were compared to one another, their growth rings matched exactly in terms of curvature, number, and width. The most irrefutable pieces of evidence were the four nail holes found in the ladder rail. When the rail was repositioned as a floorboard and four nails were inserted into the holes, the size, spacing, angle, and depth of the nails fit *exactly*. According to Koehler's calculations, the chances of this exact set of circumstances occurring twice were one in 10 quadrillion. In contrast, the DNA fingerprinting of blood found on the driveway during the O. J. Simpson trial produced a much less certain match of 1 in 170 million.

There were other factors. The ripple effect of the knots transferred across both boards. There was a pile of sawdust below the end of the floorboard where it had been sawn off while in place. The planer marks on the boards matched. Hauptmann was a dead man, figuratively and literally. On April 3, 1936, he was executed in the electric chair.

The saga of the Lindbergh kidnapping didn't necessarily end with 2,400 volts. Mark Falzini, an archivist who oversees the New Jersey State Police museum, where evidence from the kidnapping is kept, explained in a recent interview that he was aware of at least sixteen people—including a black woman from Trenton— who claimed to be Lindbergh's kidnapped son.[19]

No other forensic wood investigation is as famed as the Lindbergh case, and few involve such cut-and-dried evidence. Alex Wiedenhoeft, a botanist at the Forest Products Laboratory, became so fed up with receiving botched and tainted wood samples sent for analysis from crime scenes that he wrote an article titled "Wood Evidence: Proper Collection, Documentation and Storage of Wood Evidence from a Crime Scene" for *Evidence Technology Magazine*.

The motivation for writing the article came from a case where a body was discovered partially buried by three rotten logs. Detectives at the scene of the crime collected a huge plastic bag of rotten wood from one of the logs but no bark, no wood from the other two logs, and no wood that wasn't rotten. A suspect was arrested, and lo and behold, wood debris was found in his pant cuffs. But there were two

Alex Wiedenhoeft of the Forest Products Laboratory stands behind a crumpled section of a small plane, holding a fragment of the wood propeller that helped determine the cause of the accident.

problems. The debris found in the cuffs was bark material, and no bark material had been collected from the crime scene. Secondly, by time new investigators went back to gather more bark material, the area had flooded, and much of the debris had moved around. The detectives still gathered up wet bark and wood samples and sent them to Wiedenhoeft in sealed paint cans. "And when I opened the paint cans I got clouds of black mold spores all over everything." The wet material had turned moldy in transit and had become useless for identification. Wiedenhoeft maintains that if the first detectives had gathered the evidence right in the first place, they could have had their guy.

Wiedenhoeft points to a crumpled mass of metal on the table as an example of correct collection of wood evidence. Two small airplanes, one with a wood propeller, were involved in a midair crash, and there was uncertainty about how the accident was caused. Analysis of wood embedded in the metal of one of the planes showed it to be Sitka spruce, a wood with a high strength-to-weight ratio often used to manufacture propellers. From this evidence, it was determined who ran into whom.

In another aviation-related case, the FPL was called in to settle conflicting reports on how a crash occurred. Some witnesses said smoke was pouring out of the engines as the plane went down; others said the engines were on fire; others said the engines weren't running at all. Plant matter taken from the insides of the engines was analyzed. The discovery that the wood and twigs were ground up proved that the engines were running when the plane hit the ground. And since the material was charred, not burned, even the approximate temperature of the engines upon impact could be determined.

In the 1920s, 1930s, and 1940s, the Forest Products Laboratory had six scientists performing as many as eighteen thousand wood identifications a year. During World War II, when many planes and boats were still made of wood, three shifts worked 24/7. When asked about the types of identifications done today—the annual number has dropped to about two thousand—Wiedenhoeft explains, "What's common is the breadth of weirdness." A few weeks before my visit, he'd done wood analyses for a graduate student working on a theory that a person's social status could be determined based on what type of wood the coffin was made of. He had just finished a hundred and thirty wood identifications to help in the authentic restoration of the United Nations headquarters in New York City. He and others have examined wood specimens from King Tut's tomb, the Le Brea tar pits, Blackbeard's ship, and a site at a high elevation in Turkey that some people maintain is the place where Noah's ark landed. Another recent project involved studies aimed at increasing the durability of the wood propellers used on unmanned drones in Iraq. During my visit he receives two small packets from Egypt containing wood samples from artifacts believed to be twenty-five hundred years old.

He points to a row of boards lined up against a wall and explains that he's just completed identifying woods from a recent container full of hardwoods to determine whether or not they have been legally imported under the CITES agreement. They were legal. Forensic botany marches on.

PENS AND PENCILS: GETTING TO THE POINT

I sit here effortlessly pecking away at the keys of my laptop, unappreciative of the trials and tribulations endured by writers before me. Writing, whether in the guise of literature, letters, or bookkeeping, has been until fairly recently a messy, smudgy, smeary affair. And wood has always been in the thick of things.

When the Sumerians developed their cuneiform writing system in 3000 BC, they pressed symbols into clay tablets using pointed sticks. Early movable type was made of wood. And the paper we write on is little more than wood fiber that's been—excuse the expression—beaten to a pulp.

Wood in its rawest form was even used for bookkeeping. Up into the 1820s, English tax collectors would hand out willow twigs to citizens paying their taxes. The branch was first notched to indicate the amount of tax paid, then split lengthwise. One half was then given to the taxpayer, and the other half kept as a record. If one needed to show proof of payment, the two sides were rejuxtaposed, and the notches, bark, and grain needed to match exactly. It was a receipt that made counterfeiting and tax dodging impossible. Legend has it when the system was discontinued and the twig files burned, the fire ran out of control and torched the House of Parliament.[20]

The most important inks in history owe their existence not so much to wood per se as to the trees they came from. The earliest formulas, dating back to 2500 BC, were mixtures consisting of soot from pine smoke and lamp oil, along with other substances, including the gelatin of donkey skin and musk.[21] Other inks were created from berries, plants, and minerals. But the classic early pen fuel—used from Roman times until the late Middle Ages—was oak gall ink. The main ingredient was derived from gallnuts: "blisters" produced by the oak tree in response to stings by egg-laying parasitic wasps. These acorn-size galls were crushed and soaked in water. This liquid, rich in tannic acid and gallic acid, was then mixed with gum arabic and iron sulfate to create the classic blue-black ink—an ink that was water-resistant, nonerasable, and permanent because it interwove itself into the paper's fibers.

It was the ink of Rembrandt, Bach, and van Gogh. The Dead Sea Scrolls were penned with it, as was the Constitution of the United States. Unfortunately, after hundreds of years the iron sulfate in some of the formulas was discovered to be reacting with the paper and literally burning away a thin layer of the substrate below, leaving a sort of halo—a tragic form of disappearing ink. Manuscripts by da Vinci, Galileo, and others have been destroyed or nearly destroyed through this reaction—a truly messy and difficult-to-reverse affair.[22]

The application of these inks was another messy affair. Penicillums—from which the words "pen" and "pencil" derive their names—were Roman writing implements consisting of hollow reeds with a tapered tuft of animal hair inserted in one end. The reed was filled with ink and squeezed when one wanted to push more ink to the tip. But it was the quill that dominated the writing world for over a thousand years, beginning in the eighth century AD. The best quills were taken from living birds in the spring, from the outer left wing feathers; the left wing was favored because the feathers curved outward and away when used by right-handers. Goose feathers were the most commonly used, although those of the swan were considered the best, and crow feathers excelled at making fine lines. A penknife—ah, that's where that term came from—was required to resharpen the tip, and an absorbent was needed for dabbing the tip to prevent smudges.

Regardless of who used pen and ink—Albrecht Durer or Bob Cratchit—there was the looming threat of drips, smears, and spills. Make a mistake and there was no erasing it. The task required constant dipping and dabbing. No wonder people searched for a more forgiving writing utensil.

The tips of charred sticks used to create sketches on cave walls were perhaps the first pencils. But the earliest conveniently used "dry" writing implements, dating back to Roman times, were sticks of lead. Rods of the substance could be wrapped in string or cloth and used in a clumsy fashion for writing and drawing. This pure lead left a mark, but not necessarily a dark one or a smearless one. Lead alloys and substitutes were concocted that lasted longer and left darker marks. But the biggest improvement—the idea that moved lead from the province of artists and writers to the average citizen—was encasing the "lead" in wood. Records show the existence of wood case pencils as early as the seventeenth century.[23]

Not just any wood would do. The wood needed to be stiff enough to support the lead inside, yet flexible enough to feel comfortable in the hand. It needed to be soft enough to be easily sharpened, yet hard enough not to splinter. It needed to be lightweight, yet dense enough to create a case small enough to nestle in one's fingers. It needed to be affordable and easily machined. And in the end, that perfect wood turned out to be cedar—red cedar in particular and the knot-free heartwood specifically. The wood had all the necessary attributes: it was strong, straight, lightweight, easy to sharpen, and easy to mill.

Initially, in the land of plenty, there was plenty of red cedar (sometimes referred to as pencil cedar), a tree that grew primarily in Florida, Georgia, and other parts of the Deep South. The best wood came from fallen old-growth red cedars that were straight-grained and had had time to dry naturally. As this commodity grew scarce, manufacturers turned to the living forests. But they didn't turn carefully; pencil makers, in quest of only the heartwood, often left 70 percent of the wood in a tree to waste.[24] As forests began dwindling, cedar "cruisers" turned to other creative means for obtaining wood, none of which included reforestation. Barns and log cabins built from cedar were purchased, disassembled, and turned into pencils. Cedar railroad ties were pencilized. Farmers with fences crafted from cedar were given new-fangled woven wire fences in exchange for their old cedar fences.[25]

By 1912, over one billion pencils were being made annually from American cedar; thirty years later, the yearly output had increased by five hundred million. Even with the grandest red cedars yielding two hundred thousand pencils, red cedar was in rapid decline. By 1920 in Tennessee, the supply of red cedar, living or dead, was nearly wiped clean.

American manufacturers, as well as those in other parts of the world that had been importing American red cedar, began experimenting with other types of wood—basswood, alder, Siberian redwood, English lime, and other types of cedar. Incense cedar, which eventually gained favor and is still used today, had the same mechanical attributes as red cedar—but it wasn't what people were accustomed to using. It was too white, so it was dyed. And since it didn't have the same "bouquet" that people associated with the peeling of a fine red cedar pencil, the wood was perfumed.

Manufacturers experimented with other types of materials. Paper and plastic casings were developed but were rejected by the public because of aesthetics; they had a different taste when nervously chewed upon, a different texture when sharpened, and a different feel when angrily snapped in two. In the end, wood has endured.

The first manufactured pencils were crafted by cutting a thin groove in a thin strip of wood, gluing several short pieces of lead or graphite rod—initially square—into the groove, planing the graphite even with the top of the groove, and then gluing a thin strip over the top to enclose the lead. This case was then planed into a round, hexagonal, or octagonal shape to fit comfortably in the hand. Pencil making was a slow affair.

The more efficient manufacturing process that eventually evolved doesn't differ greatly today from its roots. It starts with wood slats that are 7 1/4 by 2 1/2 by 1/4 inches; the length of one pencil, the width of six, and the depth of half a pencil. The slats are evenly grooved to accommodate six pieces of lead. After the lead is in place, another identical slat is glued on top. This "sandwich" is then run through revolving knives that cut three sides of the pencil on the top side and the other three on the bottom. What emerges at the end is a group of six hexagonal pencils.

As the pencil shape and manufacturing process evolved, so did the leads. The early lead and lead alloys were troublesome. The materials were expensive and difficult to obtain. Some formulas were too brittle, while others were so hard they tore the paper. Eventually the amount of lead in the lead pencil diminished to the point where today there is exactly 0 percent lead in a lead pencil. The "lead" in a lead pencil is now a concoction of graphite, clay, and other ingredients. There's not even lead in the paint covering the wood case. And a good thing it is, considering people's pencil-chewing habits.

Even in this digital age, there's at least a grain of truth to the saying "Everything begins with a pencil." Architects, designers, engineers—even some writers—feel most comfortable with a medium that can be erased or can be applied lightly as the creative process begins, darkly as the ideas gel. It's almost as if people have developed a primal attachment to the 7-inch yellow pencil: a simple dynamo that can draw a line 35 miles long, record forty-five thousand words, and sustain seventeen sharpenings without ever crashing or being accidentally deleted.[26]

Henry Petroski, in his exhaustive tribute to the pencil, muses, "Ink is the cosmetic that ideas will wear when they go out in public. Graphite is their dirty truth."[27] The two and a half billion pencils manufactured annually in the United States today attest that lots of us love the dirty truth.

A BARRELFUL OF COOPERS, KEGS, AND TRADITION

I'm not certain what Russell Karasch is going to do with the barrel in front of us, but judging from the preparations, it looks like something inflammatory. He's rolled a large propane tank that was standing in one corner of the room outside "just to be sure." He's tested the water pressure of the hose dangling in the water trough "just in case." He's asked me to stand by the double metal doors "because you never know what could happen." And he informs me that he usually wears a safety helmet. He then adds with a sly grin, "And kids: don't try this at home."

The room is dimly lit, but I can see Russell take a bucketful of oak wood shavings and dump them inside a bottomless and topless 30-gallon oak barrel standing on end. He lights a hand-held propane torch and with a quick wave of the hand sets the sawdust ablaze. The ensuing fire is impressive, but not nearly as impressive as what comes next.

I'm at the Barrel Mill in Avon, Minnesota, a small town just far enough beyond the grasp of the Twin Cities to still feel like a real small town. Before arriving, I'm really not sure whether this is going to be a ma and pa shop or a full-fledged factory, an artisan workshop or an automated plant, a scene out of the nineteenth century or one out of the twenty-first. And I discover it's all of these. A dozen men in a wide range of ages and nationalities are scattered throughout the one-story building. The atmosphere sings "family." Some of the machines have the date 1895 emblazoned on their cast-iron frames, while others are a modern-day concoction of pneumatic pistons and digital gauges. Most of the processes are a compromise between the speed and strength of the machines and the practiced eye and craftsmanship of the men operating them.

There's a saying, "There are no amateur barrel makers," and as Karasch walks me through the factory, one can understand why. "It's not brain surgery," he explains. "Yet, to become a cooper used to require a seven- or eight-year apprenticeship—the longest of any trade." Though few coopers were mathematicians, the skill required the ability to translate a theoretical barrel (solid geometry) into a series of staves (plane geometry) and then into a finished barrel (solid geometry again). One that was watertight.

The process begins with the rough staves—pallet after pallet of them, planed smooth on both faces and rough-cut to length and

Fire bursts from the top of a sawdust-filled oak barrel as Russ Karasch of the Barrel Mill adds compressed oxygen to fan the flames. The charred barrel will be used to age whiskey for three to eighteen years; then it will be shipped overseas (or be turned into a driveway planter).

width. The wood is clear, mostly Minnesota-grown white oak, a furniture-quality wood that rings up at $4 per board foot. The staves are quartersawn, and some have grain that would rival the beauty of any board found in a Gustav Stickley sideboard. For wine and whiskey barrels, white oak has always been the wood de rigueur regardless of whether the barrel was made in twenty-first-century Avon, Minnesota, or seventeenth-century Seguin Moreau, France. White oak has all of the required qualities. It's widely available, strong, and hard-wearing, yet capable of bending. Its cells contain tyloses, which keep the wood watertight while still allowing for a slow exchange of air between inner barrel and outside world. This slow evaporation of spirits that subsequently takes place is referred to as "the angel's share" in the wine world.

The staves come from Staggemeyer Stave, a company in southern Minnesota that annually saws fifty thousand logs from hundred-year-old oaks into 5 million feet worth of staves. Winemakers love Staggemeyer staves because cold Minnesota winters mean slow winter growth, which translates into tight growth rings and tight-grained staves. Staves are shipped to locales as diverse as Napa Valley, Hungary, Australia, and, of course, Avon, Minnesota. In typical Midwestern practicality, nothing is wasted; wood chips go to nearby paper mills, bark to landscapers, and sawdust to local farmers for livestock bedding.

Traditional high-end French barrel makers have always insisted that the oak they use come from one of the forests planted during the Napoleonic era to provide wood for shipbuilding. The five main forests are Allier, Limousin, Nevers, Trancais, and Vosges, and each, according to legend, produces wood with distinctive characteristics that impart a distinctive flavor and bouquet to the wine. But Karasch's staves come from far less exotic-sounding places: Black Hammer, Bee, and Bratsburg.

The stave material has been allowed to air-dry outdoors for two to three years before being brought into the Barrel Mill drying room, where it's brought down (or up) to a uniform 12 percent moisture content. The staves are fed one by one into a Rube Goldberg saw/stave jointer that, in two swipes, cuts each stave to a cigarlike shape with a slight bevel along each edge. Karasch shows me a picture of this tool's predecessor, and it, like many of the old tools, is an OSHA inspector's nightmare. It consists of an unguarded saw blade, 4 feet in diameter, through which a workman runs each stave, fingers inches from the blade. The predecessor of this predecessor, which I see later, consists of an axe-shaped tool with an offset handle that coopers used to cut each stave to the right shape and angle. "The old guys were so good using this tool that they could shape a stave by eye, and they kept it so razor sharp that they rarely had to use a plane to smooth the edges," Karasch tells me, shaving a few hairs off his forearm to prove his point.

If the barrel is a "slack barrel" intended to store or haul only dry materials, the staves are tongue-and-grooved along each edge. If the

Raising rings are used to hold staves together temporarily during the first stages of barrel building. The process uses a blend of hundred-year-old cast-iron contraptions, computer-guided machines, hand labor, and mechanization.

barrel is a "tight barrel" destined for the wine or whiskey making trade, there are no tongues and grooves. As contrary as it might seem, barrel makers—at least the good ones that work within tolerances of thousandths of an inch—get a tighter-sealing barrel without the tongue and groove.

The shaped and beveled staves—a typical barrel requires twenty to twenty-five—are wheeled over to the barrel raising area. The worker sets the staves in a jig that holds them in a tight circle at the bottom and lets them splay out at the top like a Dixie cup. A pair of temporary hoops, called raising rings, are installed to draw the bottom part of the bulge tight. Next, the barrel is brought over to a mechanical noose that slowly cinches the top tighter while the worker nurses each stave into alignment. Then another raising ring is loosely slipped over that end. The process—the same basic one used for centuries—has taken one man less than three minutes.

When the barrel is constrained into its final shape, it will be called a gun. But at this point, the staves on one end are still splayed. The staves are too stiff to be forcibly bent, hydraulically or otherwise, into a completely finished gun. The worker hoists the "almost-gun" onto a slowly moving conveyor belt that passes through a long open-ended kiln. As the staves move through the heat, the lignin—the glue that holds the wood fibers together—relaxes, and the staves easily assume their new arched shape. Karasch grabs a curved stave from a rejected barrel that's been disassembled after going through the process, and he stands on it. It doesn't flex—period. "After it's cooled down, it stays this shape. You've changed its memory. It'll never flatten out again." When the hot barrels emerge from the far end of the kiln, they're often so relaxed that they lean. A practiced eye and a few bangs on the ground even up the staves so the barrel is symmetrical. In days gone by, if your finished barrel leaned, you would be ribbed about having made a "lord," as in the old English phrase, "drunk as a lord." There are no lords at the Barrel Mill.

More raising rings are forced around the barrel to draw the staves tightly together to form a gun—this time using a hundred-year-old machine that's been converted into a 5-ton press. The barrel is placed in a revolving machine—another one dating to the early 1900s—that multitasks: It trims the ends evenly, cuts the croze (the grooves near either end that will accommodate the top and bottom heads), and chamfers the lip of the barrel, known as the chime, located at the inner edges of the barrel.

And only then is the barrel brought into the dimly lit room where this story began. With the wood shavings ablaze, Karasch grabs a nozzle and hose connected to a compressor and directs a stream of air into the barrel. The fire explodes. Sparks fly and flames shoot out from not only the top of the barrel but the bottom as well. He keeps blasting air, creating a tornado of flame and a jet engine roar that makes conversation impossible. Karasch keeps peeking into the barrel, and after sixty seconds he points the nozzle away, and the firestorm sizzles to a few hot embers. He lets the barrel cool for a few minutes and carries it out to the light, where we can see the results.

Pointing to the blackened, crackled inside of the barrel, he explains, "That's what we're looking for. That's a real nice number three char. A char like that will take even raw moonshine and smooth out that barbed-wire taste in a few weeks." Charring, among other things, caramelizes the sugars in the wood, adding to the taste and color of the whiskey. But charring's not just for moonshine; all whiskeys—bourbons and Scotches—are aged in charred barrels.

Based on the expense and work involved, one might presume a fine barrel like the one before us would be used for centuries. Not so. In the 1890s the American Cooper's Union, then the strongest union in the country, pressed for, and got, a law passed mandating whiskey barrels be used just once—a law that right or wrong remains in effect today. The life span of a whiskey barrel in the United States is usually three to eighteen years. But one country's loss is another continent's gain. European whiskey and bourbon makers feel that a barrel doesn't even *start* producing its best until after two or three batches. European brewers are the main buyers of used American barrels.

Whiskey barrels are charred at this point in their production, but wine barrels are heat-treated another way—they're toasted. Through a process that takes anywhere from forty-five minutes to seventy minutes, the barrels are slowly subjected to an open flame, creating a toast that's either light, medium, medium-plus, or heavy. Kenneth Kilby, a third-generation cooper, ruminates: "Ask any cooper to tell you what constitutes the most agreeable, the most captivating smell, and he'll tell you that it is the smell of a cask immediately after it's been fired. The piping-hot oak gives off an aroma like that of a richly spiced cake being baked."[28] Different types of wines and grapes age and react more favorably to different degrees of toasting.

Think for a moment about the ingenious nature of the barrel. Consider how much more durable a wood barrel was in comparison with its predecessor, the cumbersome fragile clay amphora. A barrel could be easily rolled up and down gangplanks, streets, and ramps. No wheeled dollies or carts were required, since barrels contained their own built-in mode of transportation. Steering a barrel containing even the heaviest material could be easily accomplished by applying a wee bit of pressure with a stick or a foot. They could be stacked, standing upright or nested on their sides, five high in the hold of a ship or in a warehouse.

And strong? The arch is an exceedingly strong shape. But with its double curvature, a barrel is not simply an arch, but an arch giving strong consideration to becoming a sphere—a geometric shape with an even more impressive pedigree. Mathematicians refer to it as a segmented, truncated spheroid, but merchants referred to it as a godsend. A well-built barrel could haul half a ton of goods and withstand pressure of 30 pounds per square inch without batting a bung hole. If one wants living proof—literally—of this strength, examine how many people have survived the plunge over Niagara Falls in a crate, kayak, or jet ski (yes, all have been attempted) versus a barrel.

Once the barrel has cooled from its toasting or charring, it's placed in yet another cast-iron press that partially removes the outermost raising ring so the staves spread ever so slightly near the rim. Then a circular head (top or bottom) is carefully positioned into the grooved croze, and the barrel is tightened up again. This is repeated on both ends.

The heads themselves are small works of ingenuity. In order to create panels large enough to make the large circular heads, several boards must be joined edge to edge. But no glues can be used, since glue could taint the wines or whiskeys destined to go within. How does one create a wide, watertight plank without glue? Karasch knows. Dowels are driven part way into undersized holes drilled along the edges of individual boards, cattail reeds or river rushes are inserted between the boards (a process known as flagging), and a hydraulic press forces the boards together. When wet, the reeds will swell to ten times their normal thickness—a volume large enough to seal the seams to create a watertight head. Another unexpected material used to create a watertight seal is a bread dough substance that's often injected with a jerky gun into the croze where the barrelhead sits. This "dough," too, will swell slightly when wet, sealing that seam.

The barrel is then sanded, and the temporary raising rings are replaced, one by one, by the permanent head hoops, quarter-hoops, and bilge hoops. They're tacked in place with a few short nails or screws—the only fasteners used in the entire process. These metal hoops are pre-formed using one vintage machine that cuts the banding to length, another machine built in 1914 that rivets the ends together to create a loop, and another machine from 1905 that takes the hoop and stretches one edge so it has a slight camber to accommodate the bulge in the barrel. All three machines—with no guards in sight and operating like jackhammers on meth—would unquestionably be voted "Most likely to amputate a digit" in high school. But Karasch says—with no small amount of pride—that in the nearly one hundred years the machines have operated, there have been zero recorded mishaps. When he explains that the Barrel Mill is gradually replacing its equipment by replacing "the most dangerous stuff first," I can only imagine the guillotine nature of those machines that have been retired.

The bung hole is then bored. At various times and depending on the barrel's intended use, the edge of the bung hole would be seared in order to seal end-grain pores that might affect flavor or fitted with a metal bushing or simply left alone. The barrel is tested for water tightness by filling it partially with warm water and tapping in a temporary silicone plug. As the warm water cools, a pressure differential is created, so that when the plug is pulled, a barrel, if watertight, will "whoosh."

Some barrels are for display purposes—the type of thing you might find near the cash register at Cracker Barrel Restaurants. Others are built as movie props; the barrels detonated in *Pirates of the Caribbean II* in an attempt to kill the "Davey Jones" were Barrel Mill barrels. As was the 250-gallon beer barrel featured in *Beer Fest.* But most are used in the wine and whiskey industry. The Barrel Mill supplies barrels for Robert Mondavi, Kendall-Jackson, Silverado, and most of the wine growers in Minnesota.

Once the barrel is finished, the price tag is slapped on and the process is finished—except for the protests of those who will tell us we've ignored nuances galore.

The interaction between oak and the whiskeys and wines aged within—especially the wines—is complex. Wine aficionados, wood scientists, and barrel makers will engage in *Jerry Springer*–like verbal hair pulling over the matter. There is general agreement that toasting the barrels indeed improves the flavor, but beyond this accord, the chairs start flying.

For decades, American-made barrels, built primarily for whiskey distillers, were pooh-poohed by the wine industry. As Ron Schrieve, a vintner with Beringer Wines, puts it, "Winemakers are a pretty persnickety bunch. There was always a big chasm between the bourbon and wine business—winemakers wouldn't be caught dead using the same vulgar wood."[29]

Many felt that only French oak barrels had the proper qualities for aging wine, and to some extent that was true. Initially, barrels made from American oak imparted too much influence on the wine. But eventually, people realized that it wasn't so much the wood as how it was prepared and dried.

The French created their staves by splitting the wood, while Americans created theirs by sawing. One might think that a stave is a stave, but not so. Wood is basically a mass of long slender cells packed together side by side like a fistful of straws. If one cleaves the wood along natural lines, only the sides of these straws are exposed along the length and face of the board. But when one cuts wood, the likelihood of cutting across the grain and exposing the ends of the straws along the length of the board increases. And it is through this end grain, now exposed to the wine, that the straws release too many of the chemicals within, which can create a bitter taste.

The other factor was the drying. In typical hustle-bustle fashion, Americans kiln-dried their staves instead of letting them air dry for a year or two or three as the French did. Mother Nature, fresh air, and time were better at removing sap, tannins, and bitter chemicals than was the kiln.[30] But things have changed. Many American barrel manufacturers now use the French methods of splitting and aging their staves.

But the wood controversy does not end here. There are arguments over whether a wood stave with wide growth rings will impart flavor at a different rate than one with tight growth rings, whether red wines age better in French oak barrels and white wines age better in American white oak, whether Burgundy is best aged in stubby barrels and Bordeaux in barrels with a longer curvature.

There are few absolute truths; taste is subjective. Scientists have chemically analyzed white oak and tell us that the volatile phenols in the oak contribute to the vanilla taste; furfural yields sweet and tasty aromas; lactones impart a woody aroma; terpines in the wood provide "tea" and "tobacco" nuances; hydrolyzable tannins drive the "mouth feel" of the wine. And many maintain that every species of oak contains these ingredients in different amounts, thus affecting the wine in different ways.[31] But determining the degree to which wine is affected by oaks of different species, grown in different environments and treated in different manners, is where the conjecture lies. For under the microscope, all species of white oak are anatomically identical. And there are dozens of white oaks: swamp white oak,

Oregon white oak, overcup oak, chestnut oak, post oak, and European oak.

Alex Wiedenhoeft of the Forest Products Laboratory explains that wood, at heart, is "evolutionarily conservative." As trees evolve, the seeds, leaves, and bark may notably change, giving birth to new species, but the wood itself evolves at a glacial pace. He maintains that while there are indeed some differences between the various species of white oak, you'll find just as much variation among trees of a single species of white oak or even within a single tree. Environment, growth rate, minute genetic differences, and exposure to trauma will all affect the physical and chemical makeup of an oak, or any other type of tree.

Yet, you'll never convince a serious wine drinker or winemaker that specific species of oak don't leave their signature. Karasch talks about a gentleman who, in a blind taste test, could tell not only the difference between a wine aged in French oak and one aged in American oak, but whether the oak came from Pennsylvania, Ohio, or Illinois.

Which all adds to the wonderfully complex antics vintners go through to age a wine. Most large winemaking establishments have a keg master, who will select a barrel of a particular type of oak, based on the grape and desired outcome. If a grape has a good ripe fruit character but not much spiciness, a high-end vintner might use a barrel made by Francois Freres in the Burgundy area of France. For wine that lacks length of finish, he might use a barrel from Taransaud, a French cooperage with a lineage dating back nearly three hundred and fifty years.[32] The cooperage of Seguin Moreau makes barrels of Limousin oak, renowned for imparting a rich vanillalike flavor to the cognac stored within. Remy Martin uses these barrels to produce their Grand Cru, which can easily sell for $1,500 per bottle. If you do the math you can see that a single barrel can contain nearly a quarter of a million dollars worth of cognac.

As one might surmise, since the oak is giving, giving, giving to the wine, eventually it has little left to give. Most wine barrels are retired after five to seven years and then condemned to being cut in half, filled with dirt, and planted with petunias along suburban driveways. This having been said, one can go to France and visit wine cellars where the same barrels have been doing their job for over a hundred years. In such cases, these wise, mature barrels are contributing a mellow atmosphere in which the wine can age rather than imparting particular extractives. The other variable imparted by oak is a slow controlled exchange of air. This oxidation results in decreased bite, richer color, and more stability But, one must not go too far in giving credit to the wood and barrel instead of the grape. Randall Grahm, winemaker at Bonny Doon Winery in California, reminds us "Oak is a condiment, not a nutrient. People who think of oak as a primary flavor in wine also tend to think of ketchup as a vegetable."[33]

Up through 1890, barrels could be, and were, used for hauling nearly everything: gunpowder, nails, whale oil, water, whiskey, flour, apples, salted fish, wine, crackers, butter, and dead admirals. A vintage publication containing a "Tell-at-a-Glance Index" from Northern Cooperage—whose machinery is now gainfully employed at the Barrel Mill—reveals the wide range of barrels and woods and linings used sixty years ago. Candy syrup was shipped in Douglas fir barrels lined with paraffin, lard in basswood barrels lined with silicate, lacquers in "tight sap white oak" lined with glue, and gums in gumwood with no coatings. Red oak, basswood, ash, and spruce were also used.[34]

Barrels were vital for trade. The 1870 census reported there were nearly five thousand cooperages in the United States employing nearly twenty-five thousand coopers; only a quarter as many people were employed in the box building trades. The United Kingdom produced one million barrels a year for herring alone, up into the 1940s.[35] "Every ship in the Navy, up through World War II, had at least one cooper on board, and an aircraft carrier might employ four or more," says Karasch. Coopers and cooperages are a much rarer breed these days. "There are only twenty-two companies in the United States today with a capacity of over ten barrels a day," he says.

One barrel-making company is in an orbit of its own: The Independent Stave Company of Lebanon, Missouri, claims the title of

World's Largest Barrel Factory, churning out a million wine barrels and 750,000 whiskey casks per year. Although a fair amount of hand labor is still involved, they've pushed the craft into the digital age. They've developed chemical and sensory profiles for each type and size of barrel, with the toasting process controlled by touch screen monitors. The age of their seasoned staves is audited and verified by the firm of Ernst and Young.

Wooden kegs and barrels have a long, industrious history. A wall painting from an Egyptian tomb dating back to 2690 BC shows a wooden tub with staves held together with wood hoops. The Greek historian Herodotus writes about casks made of palm wood used to transport goods downriver to Babylon. The Bible makes reference to barrels,[36] as did Caesar Augustus.

Barrels have their dark side. The Roman barbarian Maximin allegedly used barrels during his siege of Aquileia by bringing massive amounts of wine for his troops in 54-gallon wood casks, then using the empty barrels as pontoons to cross the river in to attack. Marcus Atilius Regulus, a Roman commander who broke his word to the Carthaginians after his defeat, was taught a lesson by being rolled to death in a spike-studded barrel. Sir Francis Drake, in a successful attempt to stymie an attack by the Spanish Armada in 1588, burned the supply of staves the Spanish had set aside for building barrels for transporting water and supplies for their attack. The Spaniards were forced to make barrels out of green oak, which leaked, rotted the food, and spoiled the water.

In their heyday, barrels were in such common use that coopers specialized. Dry coopers made containers in over a hundred different shapes and sizes. Drytight coopers specialized in barrels that kept out moisture for items like gunpowder and flour. Wet coopers made casks for keeping in liquids; they too had over a hundred different designs in their arsenal. White coopers made straight-walled containers like buckets and churns, and even this specialty group had thirty-five sizes and types. Coopers made casks, barrels, buckets, butter churns, tubs, hogsheads, firkins, tierces, tuns, and butts, each with its own distinct shape and capacity. A tun with a 252-gallon capacity, weighing a hair over 2,000 pounds, gives us the term "ton." The world's largest barrel is from the 1600s; it is 22 feet in diameter, with a capacity of 50,000 gallons. It's so large that it required thirty oak trees to build—two staves per tree—and has a dance floor on top of it.

The barrel-making industry was caught in a catch-22 during the late part of the Industrial Revolution. The same technological advances that allowed the barrel industry to mechanize were also allowing the makers of other types of vessels to mechanize. Vessels like metal cans and bottles would eventually throw the barrel-making industry into decline. Though machinery was developed that could cut the time it took to make a barrel in half, the demand was eroding even faster. The forklift, making its debut in the early 1900s, suddenly produced more cheaply made boxes and crates that were easier to move, load, and stack.[37] Companies found themselves in a toss-up as to whether it was more cost effective to mechanize their barrel-making shops at a high initial outlay or to keep building their barrels manually, which in the short run cost less. And that's where the industry has generally remained: partly mechanized, partly humanized.

At the end of the morning, we drive to Karasch's home workshop, where he has a collection of old cooper's tools; the most treasured is from John Jackeloni, considered by many to be the world's best hand cooper. Karasch starts pulling vintage tools out of a drawer, many with blades wrapped in cloth to protect the razor-sharp edge. He pulls out hoop drivers for installing hoops, froes for hand splitting staves, concave and convex planes for shaping the insides and outsides of barrels, tools for inserting cattails between staves and hammers. These are tools that could have been found (and actually *have* been found) in an eighth-century shipwreck.

"Jackeloni could make any shape barrel you wanted," says Karasch. He made them diamond shaped, heart shaped, and hourglass shaped; one has an oval on the top going one way and an oval on the bottom going perpendicular to it. He could make little dinky ones that would hold a quart and others the shape of footballs. He was an absolute master.

Jackeloni had five sons, none of whom wanted to go into coopering. "We heard he was going to stop coopering and get rid of his hand tools. When we went to see him, he visited with us for two or three hours before he'd even *show* us the tools," says Karasch. "It was like we were adopting them. He wanted to make sure I wasn't going to just sell them at an auction or flea market, but use them. And that when I got done using them, I'd pass them on to someone else who would use them."

And so goes the modern-ancient tradition of coopering—a trade wherein the tools have changed a lot, the process a little, and the pride in workmanship hardly at all.

TRUE RELICS OF THE CROSS

During the 1200s and 1300s, rulers and religious figures were struck with such "true relics" mania that one observer was led to comment that if all the true relics of the cross were assembled in one place, they would be "comparable in bulk to a battleship."[38] Why all the fuss? If you had a true relic you were special—and so was your church, and so was your city, and so were your coffers. True relics made for a tremendous tourist attraction. People went to no ends to obtain a relic—or to fabricate their own relic with a story to match.

Regardless of religious beliefs, tracing the history of such a powerful symbol is a fascinating endeavor, though one inherently fraught with conjecture, dead ends, and guesswork. One common chronology of events maintains that immediately after the crucifixion, the cross, crown of thorns, and other relics stayed in the general vicinity of where they were employed. The Roman emperor Hadrian laid waste to Jerusalem early in the second century AD and built his own structures, including a temple of Venus, on the ruins. Some say that this temple was built so Christian pilgrims worshipping on this site would appear to be worshipping Venus.

Around 327, after Constantine, the first Christian Roman emperor, had made peace with the Church, excavations began in a quest to discover and unearth two holy sites in particular: the hill of Calvary, where the crucifixion took place, and the holy sepulcher, where Christ was entombed. According to some versions of the story, it was during these excavations that the cross was found. History has it that the Empress Helena—Constantine's mom, then a woman in her seventies—led the contingent that found the surviving fragments, referred to as the *lignum crucis*, or wood cross.[39]

Stories and theories regarding the discovery abound. One maintains that three crosses were found: that of Christ as well as the two of the men crucified on either side. In an attempt to determine which was the true cross, each was brought before a terminally ill woman. First cross: no reaction. Second cross: ho hum. Third cross: bingo, instant health. The explanation for how wood fragments exposed to the elements could have survived for three hundred years is often based on where the fragments were found. Some maintain that the crosses were disposed of by throwing them into a cistern, an anaerobic environment that can preserve wood for thousands of years.

It's worth noting that wood in the Jerusalem area was, and still is, scarce and that—contrary to most artistic portrayals of the moments leading up to the crucifixion—Christ most likely carried only the horizontal bar of his cross. The vertical member of most crosses was reused over and over again. That of Christ's cross was most likely used and reused for another ten or twelve years, when the city was redistricted and the site of executions was moved farther toward the city's outskirts. So it is the horizontal member that is the most unique to Christ and "the truest" of the true relics—probably.

Nonetheless, Helena was certain she had discovered the real things. The Santa Croce in Gerusalemme, a basilica in Rome, was built in the 340s AD to house some of the relics that were found. Today, a side chapel houses (again arguably) a part of the Titulus Crucis (the panel above the cross inscribed "king of the Jews" in three languages), two thorns of the crown, part of a nail, and three wooden pieces of the true cross. Along with these fragments, one can find one of St. Thomas's fingers and fragments from the grotto of Bethlehem. Helena took two of the nails used to fasten Christ to the cross and, being the overprotective mom, sent them to Constantinople, where they were forged into a helmet to protect Constantine in battle and a horse bridle to protect his steed.

Other relics, including the bulk of the cross, were left in Jerusalem in a newly built basilica. One written account, dating to 380, describes the Good Friday ceremony involved in the veneration of the cross. The relics were removed from their silver-gilt casket and placed on a table. Then, "The bishop stretches out his hand over the holy relic, and the deacons keep watch with him while the faithful and catechumens defile, one by one, before the table, bow, and kiss the Cross—. This minute watchfulness was not unnecessary, for it has been told how one day one of the faithful, making as though to kiss the cross, was so unscrupulous as to bite off a piece of it, which he carried off as a relic."[40]

In 614, Sassanid Khosrau II captured Jerusalem and removed the cross relics as a trophy. A few years later, the Emperor of the East, Heraclius, defeated Khosrau and recaptured the relics, which eventually wound up back in Jerusalem. Around 1009, Christians in Jerusalem hid the relics, and they were rediscovered during the First Crusade ninety years later by Arnulf Malecorne, who carried them with him as he led his troops into battle. They were captured by Saladin in 1187 and then disappeared.

This is just one of the tales of the relics. Other tales have the fragments carried in reliquaries worn as necklaces, carved up by bishops and given to knights, who in turn donated them to churches that incorporated the fragments into altars. There's evidence of fragments reaching Africa, Syria, Italy, and other places.[41]

Sainte Chapelle in Paris was completed by Louis IX in 1248 to house two relics of the cross "as large as the leg of a man"[42] as well as the crown of thorns he acquired from Constantinople. This project in part was prompted not only so Louis would be considered the "most Christian" king but also so Paris would be considered the new holy land. The chapel, which stands today, is nothing short of spectacular, with 6,500 square feet of stained glass windows, unrivaled in terms of quantity and quality, and seemingly supported by thin air. The panes portray the history of Christianity from creation on. And while it cost 40,000 livres to build the fabulous chapel, that amount pales in comparison with the astronomical 135,000 livres Louis paid for the crown of thorns alone. Unfortunately, during the French Revolution, the chapel, viewed as a symbol of royalty, was damaged and the relics were scattered. At one point the chapel was even denigrated to the status of flour silo.

Determining the type of wood the cross was made from is just as speculative as determining which pieces are real. One legend says it was made from aspen, and therefore the leaves of the aspen still tremble today. Others claim it was made from mistletoe, maintaining mistletoe was once a full-fledged tree but, after being used for the cross, was condemned to be a creeping parasite. Another legend talks about how all the trees but one in the forest agreed to shatter themselves so they could not be used for the crucifixion. The one tree dissenting—the holm-oak, or ilex—was the only nonshattered tree the soldiers could find. And in some places the ilex is still superstitiously avoided.[43]

Four different relics from four different churches were recently microscopically examined, and all the pieces were found to be of olive wood. Whether this unanimity was based on true relics is unknown. In 1870, Rohault de Fleury added up the volume of all the known relics of the cross and compared that with the volume of a cross 10 to 12 feet tall and 6 1/2 feet wide. He determined that of the 6 1/3 cubic feet in the original cross, only 1/6 cubic foot could be accounted for, with over 6 cubic feet lost, destroyed, or unaccounted for—a far cry from a battleship's worth. His microscopic examination of the fragments led him to determine that the cross was made of pine.

In *The Golden Legend,* written by Jacobus da Voragine in the thirteenth century, he asserts that the cross was made from four types of wood: palmwood, cedar, cypress, and olivewood.[44] Another legend says that it was made from dogwood, a tree that back then grew so straight and strong it was the logical wood of choice for Christ's crucifixion. The tree was so anguished by this state of affairs that Christ promised the dogwood it would thereafter grow as a small twisted tree, never again to be used as a cross.[45] The dogwood legend continues with the interpretation that on the outer edge of each petal there are images of rusted nails with blood on them and that the center of the flower contains a pattern resembling a crown of thorns. The primary refutation of this notion is that dogwood of this description grows only in the eastern part of North America.[46] To shine an even brighter light on the wood and bring the story full circle, some legends maintain that the cross was crafted from a tree grown from seed cast on the soil of Eden.

The cross is not the only true wood relic that people have aspired to own. In his 1973 bestseller *The Spear of Destiny,* Trevor Ravenscroft maintains that Hitler started World War II in order to capture the spear that pierced Christ's side while he was on the cross; when the spear was captured by American troops, Hitler committed suicide to fulfill the legend that losing the spear meant death.[47]

FIFTY BILLION TOOTHPICKS CAN'T BE WRONG

In the argument over which event truly marks mankind's transformation from primitive savage into rational thinker, forget about the invention of the wheel, the taming of fire, and the "rock cracking open the coconut" theory. If you want to touch and feel the symbol that represents the point when civilization became civilized, reach into the front of your kitchen junk drawer and pull out that 750-count box of Diamond toothpicks. For as Christy Turner, an anthropologist at Arizona State University, states, "As far as can be empirically documented, the oldest demonstrable human habit is picking one's teeth."[48] Indeed, Neanderthal skulls have been unearthed that show signs of teeth being picked by some kind of tool.

Toothpicks haven't been around as long as teeth, but they have been around for millennia. The Greeks and Romans were fond of porcupine quill toothpicks. But wood has been the norm throughout history, whether it was wood from the neem tree in India, the spice bush in Japan, or bamboo, cedar, and willow in China. The prophet Muhammad used toothpicks so regularly that he had a servant designated as master of the toothpick who carried toothpicks made of aromatic aloe wood dipped into the holy water fountain at Mecca. There are records of toothpicks being used in Japan as early as AD 538, and one source—though difficult to verify—maintains the toothpick was popularized by Buddhists as being one of the "eighteen essential items."[49]

The toothpick was an integral part of early etiquette. In Hugh Rhode's *Boke of Nature,* printed in 1577, he admonished, "Pick not they teeth with they knife nor finger-end. But with a stick or some clean thing, then do ye not offend."[50] The European aristocracy loved their toothpicks. They developed exquisite versions crafted from silver, gold, and ivory and often wore them on neck chains. The French served them at the table, originally by sticking them into the dessert but later with a plate and napkin at each setting.

Yet, toothpicks were not to be denied to the common man. In 1869, Charles Forster, who worked for a wooden shoe peg manufacturer, developed a machine capable of mass-producing toothpicks. He used white birch from Maine, since it was white, easy to work with, and left no after-taste. It was, as it turns out, also an excellent choice because some research indicates that the chemicals within birch wood can help block the bacteria that lead to tooth decay.

Forster used a clever marketing gimmick for jump-starting sales. Since few restaurants in Boston offered toothpicks, he spurred demand by paying several Harvard scholars to request them after dinner; if the restaurant had none available, the students would protest loudly and threaten to never visit the restaurant again. When this was done at the Union Oyster House, the restaurant began buying Forster's toothpicks, and many other restaurants soon followed suit.

Forster's Inc. still cranks them out in Strong, Maine, the "toothpick capital of the world," where ten workers produce twenty million toothpicks daily—an output approaching fifty billion toothpicks a year: nearly ten for every person on the planet. That yearly production, according to Richard Campbell of the company, is enough that—laid end to end—the toothpicks would circle the world thirty times. If you're curious about the raw materials involved, a cord of wood will yield 7,500,000 toothpicks.[51] Forster's was bought in 1947 by Diamond Match—a company that knows about small wooden things, since it also cranks out twelve billion wooden matches a year. The company today accounts for around 90 percent of all toothpicks sold in the United States.[52]

The manufacturing process involves steaming the wood, rotary-slicing it into toothpick-thick veneers, rough-dimensioning and shaping the toothpicks, and tumbling them against one another to smooth their surfaces. Eventually they're packaged; 750 per box for the flat ones, 250 for the round. Even when one takes into account the total mechanization of the process, it's difficult to fathom how one can make 750 of anything for 79 cents. There is at least one notable exception to the low price of toothpicks: a twenty-three-year-old toothpick taken out of the warmup jacket of former New York Mets pitcher Tom Seaver brought $440 at auction in 1992.

There are dangers even in toothpicks. According to the Consumer Product Safety Commission, nearly nine thousand toothpick-related accidents occur every year. But all in all they're fun, versatile little things.

Different cultures gravitate toward different toothpick shapes. The rounded toothpick dominates in Japan, the flat toothpick in the United States, and the triangular toothpick—well designed, since it's the same shape as the interdental space where most food debris seeks refuge—in northern Europe.

Those especially enamored with the toothpick can visit the toothpick museum in Kawachinagano, Japan, a city responsible for the production of 95 percent of Japan's toothpicks. There you can learn about the history and culture of toothpicks, view toothpicks from over fifty countries, and gaze upon rare toothpicks from around the world. On display you will also find the first imported toothpick-

Toothpick City is a conglomerate of scale replica structures (1:164) from around the globe. It currently contains over two million toothpicks; fifty-six thousand of them were used to construct the Brooklyn Bridge in the foreground.

manufacturing machine and a combination ear cleaner–toothpick—a product which, perhaps not so surprisingly, never garnered a strong following.

Some prefer using toothpicks for artistic purposes rather than dental hygiene. Joe King used 110,000 toothpicks to create a 23-foot-tall version of the Eiffel Tower, and Wayne Kusy built a 16-foot replica of the *Lusitania* using 193,000 toothpicks. But Stan Munro is in an entirely differently league. Don't call him a toothpick artist. "I'm a toothpick engineer," explains the builder of the two-million-component Toothpick City. His conglomerate city contains the towering 36,000-toothpick Sears Tower, the amazing 56,000-toothpick Brooklyn Bridge, and the 38,000-toothpick Yankee Stadium (with a seating capacity of "three, maybe four").

His initial inspiration came in sixth grade, when the assignment was to build a structure out of toothpicks sturdy enough to support a hard-boiled egg. "We kept adding books and more weight on top of my structure, and it didn't budge, so finally I turned my desk upside down and set it on top and it still didn't give way. That's when I started realizing I could get attention through toothpicks."

Later on he was motivated by the idea of breaking the Guinness world record for toothpick structures, which at the time stood at one million toothpicks. "One week before I was going to file the papers for my two-million-toothpick city, some guy comes along and registers a three-million-toothpick alligator." When he explained to the Guinness people that the alligator was built of solid toothpicks through and through—a no-no in the world of toothpick architecture etiquette, where structures are normally hollow—they told him to tip his structures upside down and fill them with toothpicks in order to claim the record. "I told them to forget it," explains Stan, "I've got my principles." Principles that have neatly paid off, since he recently sold his entire Toothpick City to the House of Katmandu in Mallorca, Spain. It's now rumored that a woman in South Carolina is set to turn the Guinness toothpick record on its head with a giant seven-million-toothpick cube.

Munro uses square-body toothpicks, which he buys by the case, and everyday Elmer's Glue-All. His saw is a scissor, and his workshop is his living room floor. "I've been very lucky that my wife is a podiatrist," he explains. "I can't tell you how many times I've had to use her services after walking barefoot through one of my toothpick construction zones."[53]

In this chapter we've looked at the playful, religious, working, and artistic sides of wood. But wood also has a deadly side. Universally available, strong yet flexible, and easy to shape, wood is the ideal material for making everything from 20-ton catapults to 2-ounce arrows.

Next are a few of those stories.

CHAPTER 8

Wood, Weapons, and War

What is it that brings forty-five thousand people to the World Championship Pumpkin Chunkin competition in Sussex County, Delaware, every year? Why has an eccentric millionaire Englishman spent years and thousands of dollars developing a catapult that will throw a Buick a quarter of a mile? Why is it that when you Google the words "catapult" and "fun" together you get 1,500,000 results? Because catapults are cool. And why are they cool? It's because within every catapult—whether it was made yesterday or two thousand years ago—lies the potential for a boneheaded move of such gargantuan proportions that you have to be fascinated by them.

TEN GREAT MOMENTS IN CATAPULT HISTORY

With the exception of early siege engines (equally cool), there has never been a larger weapon made of wood than the catapult. In the heyday of catapults, no machine of any kind was more expensive, more powerful, or more massive. Construction of the largest required strategic planning on a par with that of coordinating soldiers on the battlefield. A catapult used at the Battle of Acre needed a hundred carts to haul it. The massive War Wolf catapult required "fifty carpenters and five foremen a long time to complete."[1] A group of twenty-four catapults seized and dismantled by Louis IX provided enough timber to build a stockade around his entire camp.[2]

The weapons were forever evolving and complex. Indeed, the term "engineer" is derived from *engynour*, the title given to the men who oversaw the construction of catapults and other war machines. Sawyers were required to find and cut the straightest, strongest trees, nearly always oak; massive carts were required for hauling the timbers; cranes and gin poles were required for assembling components, beams, and levers.

Rock-solid cross-bracing was needed to stabilize a structure designed to hurl rocks the weight of a piano, by means of an arm weighing several tons, that launched its payload by hitting the ground or a crossbar at full speed. Intricate interlocking wood joints, metal plates, and wood pegs were used for making the connections. So specialized was the craft that the Carpenter's Guild was formed in 1300 to provide adequate training; it exists today and is most likely the oldest workers' organization anywhere.[3]

Bill Gurstelle, author of *The Art of the Catapult*, explains: "Catapults need to be extraordinarily overbuilt. The stress on a machine of that size is unbelievable. You need to use massive oak beams and mortise and tenon joinery. People [today] try to build them out of softwood and iron and they bust themselves apart almost immediately."[4]

Why build such monstrous machines? one might ask. The answer: to deal with monstrous fortifications. Just imagine what an attacking force was up against when encountering a fort. The structures were often perched on high ground, with walls 12 feet thick and their entrance limited to one or two doors accessed by drawbridges. Often they were surrounded by moats that contained not the stereotypical alligators but something even less savory to swim through—the refuse and waste of the occupants within. Those above you could rain down stones, arrows, spears, insults, and boiling oil with pinpoint accuracy, with little risk of life or limb to themselves. Unless you were able to starve them out—a wait-and-see game that could literally take years—or bribe someone within to, ahem, forget to lock the back door after taking out the trash (a not uncommon ploy), a well-designed fort was virtually impenetrable by standard means and weapons.

The only way to level the playing field was to (1) elevate the playing field to the height of the fort's walls, which an invading army could do with siege towers, or (2) go over, under, around, or through the walls to overcome the barrier. A wide range of catapult devices were crafted over the years: There were spring engines, which operated like gigantic crossbows; ballistas, which tossed their payload like the killing arm of a gigantic mousetrap; and trebuchets, which worked via a counterweight (and what most people think of when they think of the classic catapult).

It's difficult to divine the exact construction details and materials used in the earliest catapults. As Sir Ralph Payne-Gallwey bemoans in his classic *Projectile-Throwing Engines of the Ancients*, "With few exceptions, all the authors [Atheneus, Biton, Josephus, Vitruvius, and others] simply present us with their own ideas when they are in doubt respecting the mechanical details and performances of the engines they wish to describe." He goes on to describe how most old renderings or paintings omit details or conveniently position a soldier in front of the working mechanism rather than "portraying it correctly."[5] We do know that the throwing arms of most catapults were made of multiple layers of wood bound together—often with broad strips of wet rawhide—to create an arm flexible enough to hurl large objects great distances, yet tough enough ("hard and tight as a sheath of metal" by Payne-Gallwey's account) to withstand relentless pounding.

Here are ten far-flung stories that represent some of the highs and lows in catapult history.

1. FLYING SNAKES AND PISSED-OFF HORNETS. In 332 BC, Alexander the Great found the island city of Tyre a troublesome roadblock in his quest to gain control of that stretch of the Mediterranean coast. The city lay half a mile offshore, with fortified walls 150 feet tall in places and no land directly outside the walls from which to mount a

siege. He constructed a manmade peninsula leading out to the island, along with massive wood towers on each side equipped with massive catapults, and began firing away.

The Tyrians had some pretty impressive catapults as well, and the two sides bombarded each other relentlessly. The Tyrians, in a novel defensive move, concocted gigantic wooden spinning wheels along the tops of their fortress walls to deflect Al's incoming projectiles; some records contend they looked like gigantic paddleboat paddles, while others say they looked more like the old-fashioned table fans we were so tempted to stick our fingers in as kids.

In an attempt to gain the upper hand, the Tyrians mounted a surprise attack and sank several of Alexander's unmanned ships. But Alexander launched a bit of a surprise himself. He ordered his carpenters to build some thin-staved barrels (not too well built, please). Your run-of-the-mill conqueror might have filled these with stones, but not Al. He ordered the barrels filled with venomous snakes, then catapulted them onto the enemy ships. Many Tyrian soldiers found it an ideal time to abandon ship. Those who stayed were greeted by a second round of projectiles—hornet nests.

One cannot pin the turning of the battle, war, or history solely on a squadron of catapulted vipers and insects. But one could surmise that it went a great distance in weakening the Tyrians' resolve, not knowing what might land next.

2. LEONARDO'S AMAZING DRIBBLING MACHINE. A few years ago, a group of engineers, designers, and woodworkers re-created an 80-foot-long crossbow-type spring engine based on a 1485 Leonardo da Vinci design. The machine required 10 tons of iron, 10 tons of wood, thousands of man-hours, tens of thousands of dollars, and a television crew to build it. The builders attempted to stay true to the drawings and materials that would have been available during Leonardo's time. They used a laminated combination of hickory (with good tensioning qualities) for the outside of the gigantic bow and elm (with excellent compressive qualities) for the inside. They were hoping for a distance of 200 yards.

On the momentous day of the test, with cameras rolling and grant administrators clenching their teeth, they fired their weapon. The farthest shot dribbled 60 yards—and that included the roll. As it turned out, the most dangerous part of the weapon was the weapon itself, when the giant cross arm snapped during the testing. To put it all in perspective, in 1990, American Randy Barnes threw a shot put of comparable weight nearly half that distance, using a machine that was much more accurate, was less prone to malfunctions, and weighed 40,000 pounds less—his arm.

After the gigantic crossbow experiment was over, the designers realized that if they had constructed the removable braces da Vinci had included in his drawings, the bow could have flexed more and shot greater distances.[6]

3. CREATIVE AMMUNITION. Catapult ammunition wasn't limited to mere rocks. Dead animals were often flung into fortressed cities under lengthy attacks in order to spread disease; pigs were preferred because of their superior aerodynamics. One time a messenger, who happened to be the son of royalty, was undiplomatically catapulted back into his fortress after negotiations broke down. Some people surmise that the Black Death of the 1300s became widespread after early plague victims were catapulted into a castle in the Ukraine under siege by Tartars. Many in the castle who were exposed to the plague escaped, rapidly spreading the disease.

If wooden buildings or barriers needed to be destroyed, concoctions of sulfur, sawdust, tree pitch, and oil were blended in barrels, lit, and then launched. The stickier, the fouler the odor, the harder to extinguish, the better. Iron arrows, the severed heads of messengers with notes attached, asphyxiating gases—anything was fair game. Two thousand cartloads of animal manure were flung during the siege of Carolstein. Lead balls crafted from the lead roofs of area churches were the favorite projectiles of Edward I of England.

4. THE CATAPULT SUPER BOWL. Perhaps the Super Bowl of catapult warfare was fought in the twelfth century at the walled city of Acre, along the coast of what is now Israel. Richard the Lionheart and other crusading generals from Germany and France pitted their forces against the Muslim armies lead by Saladin the Great Sultan.

Both the crusading forces and the Muslims put up a ferocious fight. The crusaders constructed a catapult so massive that they named it God's Stone Thrower. They also built movable wood siege towers with equally ominous sounding nicknames: Mal Voison, meaning "bad neighbor," and Mal Cousine, meaning "bad relation." These towers stood 90 feet tall and were covered with animal skins to ward off enemy arrows, and vinegar, mud, and other substances to ward off fire. Once the towers were positioned against Acre's walls, the plan was for the crusaders to overtake the city.

So the Muslims were facing this: a world-record-holding catapult named God's Stone Thrower knocking at the door, two siege towers with titles as ominous as that of any WWF tag team inching toward the walls, and a strategist with the nickname Lionheart mapping things out. What to do, what to do? Hands, anyone? A lowly foot soldier suggested that perhaps the best defense would be the best offense. They could build their own catapult that would launch a payload so incendiary, so unquenchable, and so sticky that it would destroy the towers despite the precautions taken to fireproof them. They built the devices and shot a few decoy fire barrels in order to determine the correct trajectory. While this "failed" weapon was being mocked by the pompous crusaders, the ballistics team inside the walls began firing barrels of the real McCoy. The massive wooden towers burned, hundreds of crusaders were killed, and the Muslims fought on with renewed vigor. Eventually the city fell, but not without a fight.

5. Syracuse and Archimedes (Eureka!) Some of the earliest siege engines and catapults were developed around 215 BC, when the leaders of Syracuse hosted a sort of siege weapon convention, calling together the great engineers and builders of the day. The Romans were preparing to invade Syracuse in order to extract revenge for the Syracusians having sided with the Carthaginians in a recent conflict. The Romans didn't even know how to say the word "small"; they were heading to Syracuse with eighty ships and fifty thousand troops. The Syracusians begged Archimedes for help. While most of us think of Archimedes as the fellow who ran naked through the

streets screaming "Eureka" after formulating a milestone scientific theory in the bath (Eureka!), he also liked to dabble in amazing war machines.

Archimedes rolled up his toga sleeves and went to work. It's difficult to separate fact from fiction, but clearly he dreamed up some novel defenses—and offenses. He developed a gigantic wood crane capable of dropping 600-pound lead balls onto ships.[7] Some accounts discuss his creation of a humongous hook, suspended from a humongous wood tower, that could reach over a castle wall, pluck a ship from the water, and drop it. Would you expect anything less from the man who proclaimed, "Give me a lever and a place to stand and I will move the world." According to the records, he developed a steam-powered cannon that could shoot projectiles a quarter of a mile, as well as rapid-fire catapults. Some even credit him for a system of mirrors that could concentrate sunlight on Roman ships and ignite them. In 1973, Ioannis Sakas, a Greek engineer, built a replica of the device using seventy bronze-coated mirrors. When he aimed the contraption at a ship made of plywood at a distance of 165 feet, his target burst into flames within minutes.[8]

In the end, the Syracusians lost. To put an even more tragic end to the tale, an enemy soldier—who felt insulted when an old eccentric man, sitting and drawing mathematical diagrams in the sand, told him to go away—ran Archimedes through with his sword. Eureka.

6. THE HEARTWRENCH OF MASADA. The legendary walled city of Masada, which the Romans attacked in AD 73, had great natural defenses. Though the attacking Romans had an estimated 8,000 soldiers versus the Jews' 960 (only 200 of whom were men of fighting age), Masada was so well positioned and protected that even fifty Roman ballistas firing 24/7 couldn't bring down the walls. Masada was located high on a hill, so it was impossible to roll siege towers up to the walls to mount an attack. But the Romans were, if nothing else, persistent. They—or more correctly several thousand Jewish slaves—spent months building a 265-foot-long 100-foot-high dirt ramp up the side of the mountain—a structure still visible today.[9] To keep the ramp

in place, logs were embedded in the hill (an early form of rebar); then dirt was layered on top until it was substantial enough to enable an 80-foot-tall siege tower to be rolled up its incline. And that's exactly what several thousand soldiers proceeded to do. The Masada Jews attempted to start the tower on fire, but the Romans had sheathed it in iron. When it became clear that the Romans would lay siege to Masada the next day, 960 Jews committed mass suicide that night rather than suffer at the hands of the Roman invaders.

7. NO, YOU CAN'T SURRENDER. Some surmise that the largest single catapult, and the largest number of catapults, ever erected were for King Edward of England's siege of the strategically located Stirling Castle in Scotland in 1304. Thirteen trebuchet-style catapults were erected under the watchful eye of Edward and under the terrified sideways glances of the Scots. The Robinette, Gloucester, Segrave, and Vicar catapults were gigantic, but the largest of all was one nicknamed Ludgar the War Wolf. It took fifty-five carpenters several months to build.[10] The other gigantic trebuchets had been pummeling the castle walls for four months with limited success, but when the War Wolf was rolled into position, the Scots raised the white flag. Yet, Edward hadn't invested all that time, money, and manpower into the War Wolf only to see it gather dust. That would be no fun. No, he wanted to use it. He refused the surrender documents and cocked the War Wolf's trigger. The first 300-pound ball demolished an entire castle wall. Subsequent stones did no less harm, the castle was eventually reduced to rubble, and the Scottish defenders were reduced to thirty men.

8. ONE LAST SHOT. Around 1400, gunpowder weapons began taking favor over catapults, and by 1500, the catapults were all but kaput. One notoriously failed resurrection of the catapult occurred in 1521, when Cortez ran low on gunpowder and his solders crafted a trebuchet during their siege of the city now known as Mexico City. Either the calculations or the construction went wrong: this trebuchet launched its payload not toward Mexico City but straight up. The stone landed back on top of the trebuchet, destroying it. Sometimes you're the windshield; sometimes you're the bug.

The Warwick Castle trebuchet in action. The thirty-three-ton machine—the largest working trebuchet today—can launch a thirty-pound payload nearly a fifth of a mile.

9. CATAPULTS HURTLE INTO THE TWENTY-FIRST CENTURY. Two fellows in Texas have developed a trebuchet that can hurl pianos, toilets, cash registers, and boat motors over 100 yards. This is only a warmup in their quest to build the world's largest trebuchet with a 55,000-pound counterweight, capable of flinging a 1962 Buick a quarter of a mile. Their inspiration is an eccentric millionaire Englishman by the name of Hew Kennedy, regarded as the father

Detail of a massive catapult as depicted in the *Dictionnaire raisonne de l'architecture francaise du XIe au XVe siecle (1854–1868).*

of modern catapulting. His machines have flung over sixty pianos, a half dozen cars, and several 500-pound dead pigs, some wearing parachutes. His devices hurled pianos and coffins for episodes of *Northern Exposure* in 1992. He has a catapult named Thor that he is priming to hurl a 1962 Buick; apparently the new standard in megacatapult fodder. If he succeeds in reaching his 1,200-foot goal, his Buick will have sailed ten times the distance of the Wright brothers' first flight a hundred years ago.[11]

If you want to see what is claimed to be the world's biggest siege machine in regular action today, visit Warwick Castle near

Birmingham, England. There, once a day, you can see a 60-foot-tall trebuchet, weighing 22 tons, launch a 30-pound payload nearly 1,000 feet.

10. CATAPULTS ARE FASCINATING REGARDLESS OF THE ERA IN WHICH YOU LIVE. When I ask Bill Gurstelle why he thinks people are so enamored with catapults, he responds: "On one level they're very simple and easy to understand, but on another level they're very intricate and complex—you could base a whole physics course on how they work. But they're complex in a different way than a computer. People can look at them and go 'Hey, I understand what's going on here.'" Gurstelle would know—he's built and launched many a catapult, including his most powerful one, the War Wolf; a scaled down version of the original War Wolf. "There's something about a machine in the act of hurling that really turns people on," he says. "I know that when I show my catapult to people young or old, they're impressed with it. They just think it's fun."[12]

A TALE OF TWO WARSHIPS: ONE UNSINKABLE, ONE UNSAILABLE

Though launched nearly 170 years apart, the Swedish galleon *Vasa* and USS *Constitution,* better known as Old Ironsides, had much in common. Both were built to be premier warships of their day, carrying crews of nearly five hundred and toting fifty to sixty massive cannons each. Each sported three gargantuan masts (some as tall as 190 feet) and consumed vast amounts of oak in its construction. Both had inauspicious christenings, yet both float today. But there is one difference—a huge difference, actually. USS *Constitution,* once launched, circumnavigated the world, had an illustrious military career, traveled hundreds of thousands of miles, and floated the oceans for all of her 210 years. The *Vasa,* once launched, sailed two-thirds of a mile, listed to its port side, and—well, as you'll see, the U.S. Navy got a much better return on their investment than the Swedes.

* * *

Old Ironsides, launched in 1796, came from good stock and plenty of it. She was crafted from two thousand trees—the equivalent of 60 acres' worth of forest—and incorporated specialty woods sourced from Maine to Georgia. White pine from Maine was used for the masts, white oak from New Jersey for the framework, and live oak from Georgia and Massachusetts for the planking.

As with other massive merchant ships and warships of the day, the wood for constructing certain parts of the superstructure could only be supplied with help from Mother Nature. The wishbone-shaped wing transom was crafted from an oak tree with an unusually wide fork in the trunk, the L-shaped knees came from parts of the tree where roots or branches met the tree trunk at right angles, and the curved ribs came from trees with a natural bend to the trunk. True to the military tradition of outrageous cost overruns, she came in with a final bill of $302,700—almost $200,000 over budget—and that without the $450 toilet seats.[13]

The 4-inch oak planking—used inside and out as sheathing—created a hull nearly 2 feet thick in places.[14] This made Old Ironsides so heavy that when she was launched, her birth wasn't by natural delivery but rather by C-section. An entire dockload of dignitaries and politicians had assembled to watch the launch. On the first oomph she made it 27 feet out of dry dock, then stalled. On the second try she crept another 31 feet. Steeper launchways were built, and finally on the third try, days later, she floated.

Old Ironsides came by her nickname honestly. On August 19, 1812, she went toe to toe with the British warship HMS *Guerriere*. The two ships parried and feinted, then settled into a good old-fashioned slugfest. Within an hour, all the masts of the British ship had been destroyed, while the *Constitution* remained relatively unscathed. British cannonballs bounced off the sides of the *Constitution,* causing one British sailor to proclaim "Huzzah! Her sides are made of iron!" By the end of the battle there were seventy-nine fewer British sailors to shout "Huzzah!" and a classic nickname had been born. The ship wreaked general havoc with the Brits through the War of 1812, running blockades, sinking enemy ships, and capturing others.

Old Ironsides didn't have iron sides but a hull sheathed in a material darn close in hardness: live oak. Live oak—its name derived from the foliage that remains evergreen throughout the year—is one of the hardest, strongest woods in North America. The strategic value of ships built from live oak became so apparent that the government purchased more than a quarter million acres of live oak forestland to have in reserve for shipbuilding.[15]

Live oaks are tough customers in life, too. After Hurricane Katrina barreled through New Orleans in 2005, there were many instances where buildings collapsed but the deep-rooted, solid-trunk live oaks around them survived.[16]

Not all was smooth sailing for Old Ironsides. She lost her bowspirit's figurehead, a wood carving of Hercules, in a collision during the Barbary wars. When it was replaced by a figurehead of then-president Andrew Jackson in 1834, a merchant skipper—clearly not a great fan of Jackson—beheaded him under cover of a thunderstorm, later returning the head to the secretary of the Navy for reattachment. She ran aground near the white cliffs of Dover while returning from the Paris Exposition in 1879, and three years later she was put to anchor, disgraced when a barnlike barracks was built on her deck to house Navy recruits.

She underwent a series of restorations, the lousiest one being the first in the 1870s in preparation for celebration of the nation's centennial. In 1907, $100,000 was sunk into a minor restoration effort, but within ten years she was taking on 25 inches of water a day. By 1924, she needed constant pumping, and a restoration effort—$150,000 of it funded by children contributing pennies—commenced. In 1930 she sailed again, thanks to a nearly $1 million restoration effort. In the 1970s and 1990s, she underwent even more restoration, and again live oak was used, both for strength and for posterity.

The story of the *Vasa* is considerably different. Built to bolster Sweden's empire-building status, the 200-foot-long ship sported two decks, sixty-four cannons, ten sails, and seven hundred hand-carved wooden sculptures, many fearsome and ferocious looking to unnerve the enemy. The ship was built of wood cut from the highly protected Crown forests; one thousand oaks alone were felled to provide the timbers. Though the *Vasa* was 90 percent oak, a dozen other woods—including walnut, pear, apple, maple, and alder—were used where needed.[17] It took three years to build, and some called it a floating work of art.

The shores around Stockholm were lined with spectators for her maiden voyage on August 10, 1628. The ship set sail, traveled 1,000 meters, fired her cannons for the traditional salute, caught a light breeze, listed to one side, took in water through the open gun ports, and sank. The sinking should have come as no great surprise; in a test conducted before the launch, thirty sailors who were ordered to run from side to side nearly capsized the ship.

The hubris of King Gustavus Adolphus—who insisted that the ship, originally conceived of as a single-decker, be built as an ornate, top-heavy, double-decker—was partly to blame. But it really boiled down to lousy design. It was determined that the ship was "well-built, but badly proportioned."[18] In the 1600s, ships were designed on the basis of "reckoning," the designer's experience, and gut feelings rather than engineering studies. Recent "reckonings" show that if the ship had been built 16 inches wider and carried 240 tons of ballast instead of 120 tons, it would have floated just fine.

A futile attempt was made to raise her shortly after sinking, and a salvage operation in 1663 recovered sixty-one of the cannons. But after that, the *Vasa* remained out of sight and out of mind. In 1956, the wreckage was rediscovered and reexamined, and lo and behold there wasn't as much damage as one might expect of a ship that had been submerged 325 years. Beer barrels still containing fluid, a keg of butter, and six instrument-playing life-size carved wooden angels (along with the devil holding his ears) were among the thousands of wooden artifacts found in excellent condition.

This remarkable state of preservation was due in part to one unsavory detail: the ship had sunk directly in the path of Stockholm's sewer outlets. The fetid waters gave rise to bacteria that penetrated the wood, rendering it poisonous to fungi and rot that might normally attack the wood. The glacier-fed waters of the Baltic also helped maintain an even temperature and were so low in salt concentration that the voracious *Teredo navalis,* or shipworm, could not survive.[19] Moreover, the ship had been constructed primarily of durable oak heartwood. Indeed, the worst damage had been inflicted by the anchors of modern ships that had dragged across her.

Various ideas were presented for raising the ship, including one that involved filling the hull with ping-pong balls. It was finally raised in 1959 after cables were passed beneath the hull, and massive pontoon boats on each side winched it upward and into shallower waters in sixteen stages. Some repairs were made while the ship was still submerged. Then, in 1961, the ship was lifted to the surface. One description of the event makes it sound like the Scandinavian version of Super Bowl Sunday. "On the day the *Vasa* broke the surface, Sweden stood still. Press, radio and TV from all over the world were there. Swedish television broadcast live—something very unusual at the time. There was hardly a TV-set to be bought in Sweden anymore—they were all sold out." Children played hooky; industries ground to a halt. "It was even calm at the maternity hospitals."[20]

But there is more to restoring a ship that had lain on the ocean bottom for over three hundred years than just raising it. To prevent the wood from drying out, which would lead to rapid and irreversible damage, the hull had to be kept continuously damp. Some 4,400 gallons of water per minute were continuously sprayed on the hull. Loose parts of the ship—of which there were over thirteen thousand—were wrapped in plastic or submerged in tubs of water to thwart deterioration, and discarded bathtubs were used when containers became scarce.

Once stabilization was accomplished, the task of preserving nearly a third of a million board feet of waterlogged wood began. Polyethylene glycol (PEG), a chemical often used by turners and woodworkers on a small scale to stabilize green or wet wood, was used. PEG works by replacing the water in wood cells, so as water leaves or evaporates, the wood doesn't collapse, crack, or change dimensions. For four years a crew of five spent five hours a day spraying the inside and

The *Vasa* warship sat on the bottom of Stockholm's harbor for 333 years before being raised in 1961. Many consider it the largest single object ever preserved.

outside of the hull with a PEG solution until an automatic spraying system was finally set up.

The *Vasa* is the largest wooden object to undergo such treatment, and the process is ongoing today. In a case of "out of the frying pan, into the fire," the preservation of the *Vasa* is now meeting new challenges. The longevity of the PEG solution itself is now in question, and sulfuric acid—being created at the rate of over 220 pounds per year by the interaction of the iron fasteners and other factors with the environment—is slowly affecting the integrity of the ship's timbers.

The preservation is a case of the old intertwining with the new. While experiments with nanotechnology are used to help preserve the largest timbers, superstition and tradition are also being employed. When the restored mainmast was raised in 1993, a gold coin, following a long-running tradition, was placed beneath it for good luck. Such a coin had not been in place when the mighty *Vasa* made her first very short voyage.[21]

THE TWANG OF THE BOW

One morning in 1415, an army of five thousand English soldiers—sick from dysentery, weary from months of marching across France in search of a good fight, hungry, homesick, and drenched from heavy rainfall—awoke to find fifteen thousand fresh French troops poised and ready for battle—not necessarily what they wanted to see. At the front of the French formation stood thousands of foot soldiers armed with pikes, maces, axes, swords, and other formidable weapons. This was bad news, but what stood behind them was worse: five thousand knights clad in armor so heavy they needed to be placed in the saddle using derricks, mounted on monstrous horses bred to carry the weight of these men in full battle garb.[22] The point spread was huge that day.

The French infantry, in a mocking display of contempt, marched within 200 yards of the English troops—well out of range of any weapon of the day—did an about-face, and dropped their pants. Rather than blush, each English archer produced a peculiar-looking weapon—a bow nearly as tall as himself—and unleashed a volley of arrows. The French infantry, who seconds before had been jocularly mooning and feeling secure in being safely out of range, dropped by the hundreds. The ragtag group of English archers continued to load and fire at a rate of a dozen or so arrows per minute. The steel-tipped arrows, weighing a mere 2 to 3 ounces apiece, created a buzzing noise like possessed bees, which further terrorized the French infantry.

The mounted French knights, having seen enough, began their charge. And they, despite layers of metal armor and chain mail, met an equal fate: the arrows easily pierced the armor, the knights inside, and the horses they rode on. By the end of the day, the score read thus: French, 10,000 killed, 1,000 captured; English, 113 killed. A group of common workingmen, outnumbered three to one but well practiced in the art of the longbow, had taken on the dominant army of the day and won.

Six hundred years later and 4,200 miles away in western Wisconsin, I'm walking along a muddy path with Thomas Boehm of Ancient Archery. If it weren't for the blaze orange stocking cap and wire-rimmed glasses, Boehm could have been one of the Agincourt archers. He sports long salt-and-pepper hair and beard, wears a loose woolly smock with a belt around the waist, and totes a leather pouch on his hip. He's a Master Bowyer and—still plying his craft forty-nine years after having made his first bow at the age of fifteen—has earned the title.

Despite being a bit reclusive and shy (he feels he's had his share of the limelight playing and performing in Baroque music groups for twenty-five years), he's taken part of his afternoon to explain the craft of traditional wood bow making. We begin the conversation over a bowl of homemade soup in the 12 by 16 foot cabin he shares with his wife, Susan; continue it in his even smaller workshop just up the knoll; and finish it in the hundred-year-old barn down in the valley where he stores his wood.

Though he makes longbows like those used at Agincourt and recurve bows like those used by many hunters today, he favors the flatbow, which he describes as "efficiently stressed, more at ease with itself, and therefore longer lasting." It's based on a design that dates back at least nine thousand years and is as beautiful in repose as it is in action.

Thomas Boehm of Ancient Archery demonstrates the proper position and technique for shooting a longbow, a weapon that dates back to 1415.

Boehm employs a variety of woods, primarily hickory, Osage orange, and yew, and each must be handled and shaped differently because of its unique qualities. He runs through the construction process he uses on a flatbow of Osage orange. It starts with obtaining the right wood, which, he explains, is no easy feat. Osage orange is native only to the Red River Valley area of Texas and Oklahoma. These native trees, in competition with surrounding trees of other species to reach sunlight, grew tall and straight. However, most Osage orange found today no longer grows tall and straight. When it was discovered that Osage orange, planted and cultivated the right way, could create a living fence that was "horse high, bull strong, and hog tight in five years," ranchers began growing it on the plains to contain their livestock. Nearly 40,000 miles of low gnarly Osage orange hedgerows were planted between 1865 and 1939 in Kansas alone. Barbed wire began replacing it in 1874; yet, Osage orange—naturally rot resistant and wickedly strong—found a new role as the ideal fence post, some lasting fifty years or more.[23]

"Most of the Osage orange out there today is useless when it comes to bow making," Boehm explains. "When someone calls to tell me they have some Osage orange but it's growing on a ridge rather than in a valley, I know with 99 percent certainty there's not enough straight wood in it to make a bow." The ideal tree—the tree that makes Boehm's heartstrings twang—is 6 to 8 inches in diameter and has at least 7 feet of straight trunk with nice even growth rings. Straightness is essential because the bow stave will be split following the natural grain of the wood for maximum strength, not sawn where cross grain could introduce weak spots. For strength, the back of the bow (the side facing the target) must consist of an uninterrupted length of a single growth ring. No cross grain—not even a single cross grain—will do with Osage orange, or you're introducing an Achilles' heel.

Boehm uses an axe to split the Osage orange log into quartered sections. These sections are then set aside to dry for a year. When he's ready to build a bow, he removes the bark, then immediately coats the newly bared wood and ends with shellac to help even out the remaining drying process. The rest is a matter of gradually turning a ratty-looking split log into the polished bow that, after forty more hours of work, will sell for $600. Though he's tried to temper the demand for his bows through pricing, he still has ten months' worth of back orders.

Boehm establishes the back of the bow by paring away wood with a drawknife to reveal the darker late wood of the selected growth ring. He then begins the thinning and shaping process. He rarely uses power tools and makes limited use of even conventional hand tools. Most of the shaping is done with files, modified horse hoof rasps, draw-knives, and scrapers. The goal is to create a bow that when drawn bends in a uniform arc, with the portion below the handgrip being slightly stiffer than the portion above: a fine-tuning process called tillering. To check the tiller, the bow is positioned with its midsection balanced on the end of a gigantic wood ruler mounted to the wall. A hook attached to a rope is secured to the bowstring, and the rope is pulled through a pulley to flex the ends of the bow downward, as if the bow were ready to shoot an arrow skyward.

Boehm places his personal bow—a flatbow made of Osage orange—in the contraption to demonstrate. He pulls the string to the 10-inch mark and gradually releases it, then to the 15-inch, the 20-inch, and finally the 27-inch mark. This last is the same distance the bow will be drawn when he is actually shooting an arrow. The curve is uniform and true. He removes the bow and hands it to me. Surprisingly, upon close inspection, I can see uneven thickenings along the length of the bow. He points out little marks where thorns grew from the trunk, and since these knotlike areas have ever so slightly weakened the wood, he has left the area ever so slightly thicker to compensate. He's also laminated a strip of rawhide to the back to help reinforce the tensile strength of the

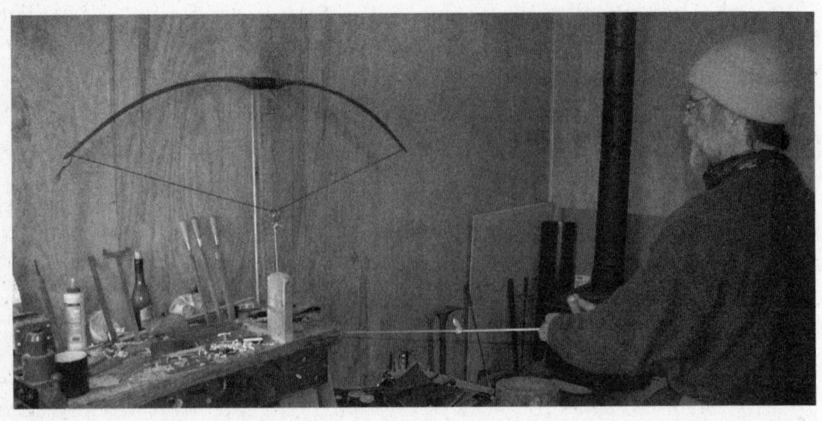

Thomas Boehm of Ancient Archery demonstrates the fine art of "tillering"—the process he uses during construction to ensure that his bows have the correct curvature and balance. Nearly all the work is done with spokeshaves, files, and small planes.

bow. This bow is the one Boehm has used to shoot fifty or more arrows every day for ten years—a total of some two hundred thousand shots. Yes, he knows how to build a bow.

When a bow is drawn, the wood on the back of the bow is stretched while the wood facing the string—the belly—is compressed. The archer is asking a single piece of wood to do two very different tasks simultaneously—to both stretch and compress—all within the small confines of the bow's thickness. When the string is released, the side facing the target snaps to contract to its original shape, while the belly "pushes" to expand. One sign of a well-built flatbow or longbow is that after years of use, when unstrung, it returns to a relatively straight shape. You want wood with "good memory." If an unstrung bow begins taking on the curve of a strung bow, it's losing its snap.

For six hundred years, longbow bowyers have understood the importance of this combination of compressive and tensile strength: that a bow's speed in returning to its straight state is what dictates the distance and velocity of the arrow. They found the ideal wood for doing this in the yew. They crafted their bows from the section of the tree where the sapwood met the heartwood. The flexible sapwood, which can easily stretch, became the back of the bow, while the denser, stiffer heartwood, which resisted compression, became the belly.

To further maximize range and speed, those who developed the longbow increased the length to match the height of the archer and created a design whereby the bow could be tensioned as far as humanly possible with the bow held in a stiffened left arm and the string drawn back beyond the right ear. This was not a weapon for wimps. A good archer of yore with a good bow could propel an arrowhead through 4 inches of brass or oak, or through seven thicknesses of boiled hide armor.

The longbow was the ancient equivalent of the machine gun; an experienced archer—and it could take ten to fifteen years to develop a good one—could shoot an arrow every five seconds a distance of 200 yards with deadly accuracy. Take an army of five thousand excellent archers, provide them with adequate ammunition, and they could unleash a blizzard of nearly sixty thousand arrows in a single minute. A formidable opponent indeed.

King Edward I so wanted to develop a corps of excellent longbowmen that he amended the Sabbath Law, which until then had permitted no other activity beyond church attendance on that day, to allow for archery contests. Henry VIII required all practice ranges to be at least 220 yards—the length of two football fields set end to end.[24]

Boehm can and will build you a traditional longbow out of yew, but he explains that it is not the most efficient type of bow because of the dynamics at work. With any bow, only a very thin strip of wood along the belly and back sustains nearly all the tensile and compressive forces. Since a longbow is D-shaped in cross-section, there is only a very small section of wood—the outer bulge of the "D" along the belly—to sustain the compressive forces. Flatbows and most other bows are thinner but wider to help better distribute those forces.

Boehm pulls some bows from a rack to demonstrate other types of bows and woods. He hands me a hickory flatbow in a partial state of completion, and explains that hickory is so strong and tough that it can be sawn from a blank without much regard to the grain. He pulls out another bow that's slightly cigar-shaped above and below the hand grip. "When most people think of American Indian bows, they think of something short. But really, only the Plains Indians, who needed something easy enough to shoot while riding horseback, used short bows. The East Coast Indians used longer bows like this." It turns out that the "short bow" is better known because few of the longer East Coast Indian bows survived. These tribes were disarmed, annihilated, or integrated much earlier than the Plains Indians, so more of the shorter bows survived.

Boehm explains that up through the 1950s, nearly all American bows were made from Cuban lemonwood. Since lemonwood—which acquired its name because of its color, not its fruit—grows year round in the tropical climate, it generates no growth rings. This, along with its great compressive and tensile strength qualities, made it the perfect wood for bows. "Then, just as I'm starting to really get into bow making in the early 1960s, the embargo goes into effect. And overnight you can't get this perfect bow wood anymore. It politicized me at a really young age," says Boehm.

He shows me a drop-dead-gorgeous, partially finished flatbow made of laminated hickory and purpleheart; then a maple bow, which he considers a failed experiment; then a recurve bow with ends that have been steam-bent into a permanent arc ready for final shaping.

"When a movie like *Braveheart* or *Lord of the Rings* or something featuring wood bows comes out, there's a big spike in orders," muses Boehm. Back when his bows sold for $200 to $300, most of his customers were reenactors of historical events. When he raised the price to $400, his clientele changed more to hunters. Now, with his bows selling for $500 to $600 and up, many of his customers are collectors. When Boehm receives an order from a customer for all three types of bows he makes, he figures he's dealing with a collector. And though large orders are nice, Boehm doesn't build bows so they can be mounted on the wall. It's more meaningful to him if they are used.

Unsurprisingly, many people who shoot with handcrafted wood bows prefer to shoot handcrafted wood arrows, and Boehm also supplies these. The lighter the arrow, the quicker the flight, so Sitka spruce, Douglas fir, and Norway pine have traditionally been used. Boehm prefers to use Port Orford cedar, a wood that grows only in a narrow band in southern Oregon. It's lightweight but has great crushing strength and strong resistance to impact bending. Additionally, it's exceedingly handsome and has a spicy perfumelike aroma.

The cedar may be hard, but it's not hard enough to resist eventual splitting when it's subjected to the repeated wedging action of the bowstring at the nock. To reinforce this area, Boehm insets a small spline of purpleheart wood running perpendicular to the nock. He also offers "footed" arrows with shafts that have a 6-inch to 8-inch section of purpleheart spliced to the tip end. This moves the balance of the arrow farther toward the tip without making the entire shaft overly heavy.

The fletching process—installing the feathers—is done by Boehm's wife, Susan. Many of the feathers come from wild turkeys Boehm himself has shot. The shaped feathers are secured with hide glue, then further reinforced with a spiral wrapping of silk thread. The fletching comes in two shapes. Swineback fletching, with the end of the feather sloping toward the nock, produces arrows that are quiet in flight and favored by hunters. The triangle cut, with the end sloping away from the nock, makes a quivering noise and is favored by those on the battlefield. Boehm states, "The fluttering noise they make when shot would, when multiplied by a flight of a thousand shafts, no doubt conjure ill in the minds of their intended targets." The French at Agincourt would have attested to that.

The finished arrow—the light red of the cedar, the inlaid purpleheart at both nock and tip, the delicacy of the silk-wound fletching—could do nothing but make one a more accurate archer. You would not want to risk losing a thing of such beauty.

Some consider the bow to be the first machine ever developed, especially if one defines "machine" as an object with moving parts that transforms muscular energy into mechanical energy. Cave paintings dating back to 5000 BC in Castellon, Spain, depict bows and arrows being used in combat; arrowheads dating back fifty thousand years from Tunisia attest to a much earlier use.[25] Artifacts reveal that the bow and arrow were used early on in Europe, Iceland, Africa, India, China, the islands of the South Pacific, and North and South America. So universally appealing was the bow-and-arrow concept that even in areas lacking proper wood, bows were crafted from animal horns or bones. It appears that the only continent where their early use went undiscovered was Australia.

The bow and arrow changed not only the way man killed, but the way he lived as well. Prior to the development of the bow, hunting and warfare were personal, close-up affairs. The rock and spear hurled from close quarters meant that every encounter was a potential life-or-death situation: kill or be killed.[26] Now a man could keep a safe distance from his prey; use cunning and skill instead of brute force to make the kill.

A good weapon never goes out of style. Ben Franklin, surely not one to begrudge innovation, had advocated the use of bows during the

Revolutionary War, pronouncing, "I would add bows and arrows as good weapons not wisely laid aside. Accuracy is as great as the musket, with the advantage of no smoke and easy procurement."[27] In *Arrows Against Steel,* John Hurley says, "No firearm approached the bow in range, rapidity of fire, accuracy, or reliability for more than five hundred years after the practical use of gunpowder in war." He goes on to maintain that "it was not until the development of the first practical breech-loading rifle, just prior to the Civil War, that a rifleman could have met the composite bow on even terms."[28] Even after firearms became dominant, archery enjoyed a brief resurgence in the Deep South after the Civil War, when the former Confederate soldiers (except officers) were not allowed to own firearms.

The bow has been used in modern warfare, too. During World War II, a colorful character by the nickname of Mad Jack—who also carried a basket-hilted sword and bagpipes into battle—favored the bow and arrow over firearms for his weapon. He silently skewered more than one sentry with the weapon. And he was not alone in this activity; dozens of German sentries were surreptitiously killed during World War II with this very ancient but very silent weapon.[29]

Despite composites, high-tech compounds, and the advantages of other types of bows, wood longbows continue to be used for both sport and hunting. Modern-day longbow makers vary in their views as to which trees make the finest bows. Some maintain that yews growing in altitudes over 3,000 feet develop better bow-making wood. Or swear that trees that are rotted on one side are better, since the non-rotted wood overcompensates by becoming stronger. Or that those of the female sex—yews are dioecious, meaning the sexes are separate—produce better woods. Or that trees scratched by bears, or those cut in the winter, are better. Whatever the selection criteria, many longbow makers feel the yew was put on earth for their use, and their use, alone. One maker maintains that one can look over a hundred trees to find one good bow stave and that "yew trees are a gift from the gods, and grown only for bows. If you see one good bow in a tree, cut it."[30]

But there's at least one yew tree you best leave alone. The Fortingall yew in Scotland is surmised to be five thousand years old, and some think it could be as old as nine thousand years. It's difficult to determine the exact age of ancient yews. They can lie dormant for years and grow at an absolutely glacial pace. The manner in which they grow also complicates accurate dating. As they age, the root system tends to pull the trunks apart, which leads to hollowing. Drooping branches work their way into these hollows and into the soil around the tree, sprouting new roots and becoming new, yet old, trees that wrap themselves around the existing trunk and grow up through the middle. The process can repeat itself indefinitely. It's rare that any given part of an ancient yew is as old as the tree itself.[31]

Though longbows are held in high regard, some think that the composite bows used by the janissaries of the Ottoman Empire were the most expertly crafted bows of all time. They were composed of a wood core, often maple, with strips of horn glued to the belly to improve compression and pieces of sinew glued to the back side for tension. Since archery was the national sport of the Ottoman Empire, meticulous distance records were kept. The longest recorded distance is 2,918 feet: well over half a mile. Other distances approaching this were recorded in stone (literally) on marble columns at an archery course near Constantinople.[32] The arrows used were featherweight, some weighing only slightly more than an ounce. The Chinese made similar composite bows using mulberry or bamboo for the core.

One other bow bears mentioning. The crossbow, which allowed the string to be kept tensioned until released by a trigger, dates back to at least 500 BC. Some employed a ratchet mechanism for drawing the string back, which gave it a range and accuracy exceeding that of many longbows. However, most were heavier, were more complicated to operate, and took longer to load; some warriors didn't like the hassle. To fire a longbow, the archer's hands needed to travel a mere 2 feet; to load, wind, and cock a crossbow, an archer's hands had to travel a total of 150 feet.[33] King Richard I of England was a great crossbow enthusiast until he was killed by one in 1199. And crossbows were expensive to build. The best were made of a wood and buffalo horn composite, but buffalo horn was an expensive commodity. And the best glue for binding the horn and wood together came from the roof of the mouth of the Volga sturgeon. The principles of supply and demand are not new; the Russian fishermen adept at catching these fish began doing some serious price gouging.[34]

There remains a solid core of wood bow enthusiasts, some opting to use the most basic and primitive of equipment. This was exemplified in the 1970s during an archery competition in England, when a frustrated observer cut a hazel branch from a nearby hedge, attached a borrowed string, and joined the match. Though competing against archers with the most sophisticated equipment of the day, this primitive archer, using modern target arrows, finished well within the middle of the pack.

But this is a bit extreme. For those wishing to shoot alongside other wood bow enthusiasts, there are other opportunities, competitions, and hunts. Thousands of archers participate in the St. George's Day longbow competition held yearly in the United Kingdom. "Self bow" and longbow competitions are common across Europe and other parts of the globe.

In wood bow competitions, the ultimate is to score a Robin Hood, a feat named after the legendary archer who legendarily split the shaft of an arrow in the target with the tip of his own arrow. As one source puts it, "Archers display their Robin Hoods as golfers display their hole-in-one balls. The arrows stuck end-to-end can be found hung with pride above mantles, next to hunting trophies or alongside letters and diplomas."[35]

And those really wishing to take a step into the past should consider combat archery; a form of medieval paintball. Conflicts can range from small skirmishes to scenarios reminiscent of Agincourt—without the death. Combatants shoot arrows with points shaped like crutch tips, and protective equipment includes modified catcher's masks, kidney belts, knee pads, and fingerless hockey gloves.

The buyers of handcrafted items often revere them in a romantic manner—and rightfully so. The custom-built guitar, the intricately inlaid pool cue, the delicately turned walnut burl vase all have an exquisite aura. But the making of these items can often be anything but romantic. The end users don't see the blisters, the failed prototypes piled in the wastebasket, the tortured balance sheets, the screaming router, the prosaic side. But Boehm's bows are indeed a romantic affair. He works with the touch and mindset of an Old World craftsman. He knows each bow stave. His spirit is in every twang.

As we walk to my truck Boehm wraps up his personal feelings, the history of his craft, and the day. "Sometimes I'll be standing there with an axe in my hand, splitting out a bow stave, and I'll think that fifty thousand years ago someone else was doing the exact same thing. Their axe probably had a flint blade, but their intents and purposes were the exact same as mine."[36]

WHITE PINES AND WAR

Today when some countries refer to national security they speak in terms of strategic oil reserves—but up until one or two hundred years ago, countries thought in terms of strategic wood reserves. If you had better woods, you had better ships; if you had better ships, you could explore farther, trade more goods, and enter naval battles with the upper hand. Access to good forests was a must.

The newly discovered Americas, both North and South, had jungles and forests chock-full of strategic woods: rot-proof South American teak for ship decking; straight Maine pines for masts; rock-hard, self-lubricating lignum vitae from Brazil for pulleys and gears. Some of the most valuable gems mined in the New World weren't gold and silver but new woods with new uses; part of expansionism was laying claim to these vital new woods.

England, in its quest for world dominance, needed to build and maintain a world-class flotilla. The hull alone of a classic seventy-four-gun English warship required 2,600 tons of timber, the equivalent of seven hundred large oak trees. England had most of the essentials for a strong navy—good ports, good sailors, and good oak for the hulls—but lacked one of the most essential raw ingredients: trees large enough to use as masts. Just as today a nuclear warhead is useless without a missile or plane to deliver it, England feared its vast war powers couldn't be delivered without adequate masts to drive its sails and ships. A state-of-the-art warship required three masts, each 120 feet tall, arrow straight, and 40 inches on the base, tapering to a minimum of 27 inches at the top. To fathom the size of such a mast, fathom this: The current champion eastern white pine listed on the National Register of Big Trees—the biggest of the species left standing—is a 132-foot-tall pine in Morrill, Maine. What is now a king was, back then, a commoner.

Since few mast-size trees remained in the English forests, the Navy had been forced into importing masts or piecing them together from Riga fir, a dense, heavy tree found primarily in Russia and the Scandinavian countries. This dependence on other countries made England queasy. When the king of Denmark halted the export of timber exceeding 22 inches in diameter in order to protect his country's own interests, England became even more jittery.[37]

But alas, the northern forests of the New World were teeming with forests of masts in the form of eastern white pines—lightweight, strong, stable white pines. Trees 150 feet tall stood arrow straight and branch free as far as the first 80 feet. Here was the solution to the mast deficit. Virgin white pine forest covered so much of New England, Pennsylvania, the Adirondacks, and other areas that the forests were nearly impenetrable. Indeed, when all was said and almost done, those white pine forests produced 750 billion board feet of lumber.[38] One source poetically describes the vastness of the early forests: "When the male flowers bloomed in these illimitable pineries, thousands of miles of forest aisle were swept with the golden smoke of this reckless fertility, and great storms of pollen were swept from the primeval shores far out to sea and to the superstitious sailor seemed to be 'raining brimstone' on the deck."[39]

In the late 1600s, the flow of masts from New England to England began in earnest, and shiploads consisting of twenty to forty masts began arriving in England on a regular basis.[40] The Brits tried establishing white pine forests on their own turf—in fact, the oft-used nickname Weymouth pine comes from Lord Weymouth's attempt to establish the trees in England—but the trees never adapted to the climate and soil conditions. And no one really wanted to wait around for the two hundred years it took to grow a good mast.

Once a good mast tree was located, no effort was spared in retrieving it. The trees, some weighing up to 40,000 pounds, were dragged or wheeled as far as 20 miles to a port. One report states that it took forty oxen to haul the largest specimens.[41]

Initially, the mast trade was a mutually beneficial situation for everyone; the Brits got their shipbuilding materials, and the colonists received a nice source of income—as much as £100 (an amount approaching $2,000 in today's currency) for a nice one.[42] But as the situation evolved into "the land of the not-so-free" things began changing. New Englanders began selling masts to some of England's adversaries, including Portugal and Spain. Around 1623, sawmills began to dot the land, and the towering white pines were converted into lumber for use both at home and abroad. There was big money to be made, and made it was. Suddenly the Brits saw their treasured masts being exported, chopped down to create tillable land, and sawn up and turned into timber frame houses. If the colonists weren't put in their place, who knows what might happen next? They might start dumping tea into a harbor somewhere. What's more, one of Britain's adversaries of the day—Holland—began capturing ships transporting loads of masts. And France, yet another adversary (did *anyone* like these guys?), disrupted the mast trade by paying Indians to harass and drive the British mast cutters out of the woods and to kill the oxen used to haul the masts.

A Parliamentary decree went out reserving all white pines greater than 24 inches in diameter for the Royal Navy.[43] The king's agents scouted out and marked the best mast trees, using three chops of the axe to create the mark of a broad arrow: the royal sign. "Thank you for locating the biggest and best trees for us," snickered the New England woodsmen as they chopped down the marked trees, cut them into short lengths, and then marked nonmastworthy nearby trees with the royal sign in contempt. In the area of Exeter, New Hampshire, alone, seventy of the seventy-one trees marked for the king's navy were "repurposed" by the colonists.[44]

When the Brits set up a spying system to determine who was felling the trees, the pioneers commenced cutting at night disguised as Indians.[45] A flustered Parliament passed another law that prohibited the felling of "any white pines, not growing within any township or the bounds, lines, and limits thereof." Never bashful about finding a loophole, the colonists in several areas divided all of their territory into gigantic townships, leaving the only available mast trees so far out in the woods that they were essentially irretrievable.[46]

In 1774, Americans stopped allowing white pine and nearly everything else to be exported to England when it became clear that those massive pines would soon be returning in the form of masts powering cannon-laden ships. In 1775, the British attacked and burned Falmouth (now Portland) in Maine because of the citizens' refusal to help deliver masts. After 1775, the Brits could be seen tooling around on the high seas with their pieced-together Riga pine masts.

Perhaps to rub a little pine pitch into the Brits' wounds, American ships began flying the Liberty Tree flag in 1775: a large white pine tree on a white background with the words AN APPEAL TO HEAVEN emblazoned across the top.[47]

PINE ROOTS VERSUS ATOMIC BOMBS

It's easy to pooh-pooh wood in an era of nuclear-powered submarines and hybrid rockets, but when push came to shove, modern-day combatants didn't hesitate to turn to wood.

Throughout 1944 and much of 1945, Japan's military commanders held fast to the belief that they could still win the war, primarily by using kamikaze pilots and battleships in the Pacific to destroy the encroaching Allied navies. However, petroleum reserves were nearly depleted. No country was shipping oil to Japan, nor was it producing any of its own. Though most dispute this notion, some accounts maintain that the reserves were so low that there was only enough fuel for Japanese bomber planes to make a one-way trip, giving birth to the kamikaze pilot.

But there was hope. Japanese chemists were way ahead of the curve in terms of creating biofuels. They had discovered a way to distill high-octane fuel from the resins in the roots of pine trees. The call went out to the Japanese populace to help in this last-ditch effort to keep Japan's war machine fueled. In response, the citizens of Japan set up an estimated 34,000 pine root stills.

This course of action came at a high price to both trees and populace. It's estimated that two and a half man-days of labor and several tons of freshly dug roots were required to create a single gallon of pine root oil. Entire mountainsides were stripped of every tree and sapling in the process. Though nearly 4 million gallons of pine root fuel were distilled per month, in the end scientists never perfected the formula to the point where it could actually be used; it left gummy deposits in the engines. Almost none of the pine root oil made it into the tanks of Japanese bombers and battleships.[48] In desperation, fuels were also concocted from coconuts, soybeans, and sweet potatoes, the last being the most successful in yielding 1 pound of ethyl alcohol per 11 pounds of tuber.[49]

Even without the pine root oil endeavor, Japan's forests were ravaged as a result of World War II. As the Japanese, both by choice and by necessity, reduced their importation of logs, they turned to their forests. The fuel shortage led to a heavy demand for charcoal, industries required heavy timbers, and the effort to rebuild bombed-out cities required massive amounts of wood. Old-growth forests were clear-cut. Even ancient revered trees—like the massive pines along the Tokaido highway leading to the Nikk Shrine—were sacrificed. Forests were cleared to plant crops for food. It's estimated that during World War II, 9 million acres of forests were lost, representing 15 percent of Japan's forests. By the end of the war, 50 square miles of forests were being clear-cut every week.[50]

Pine bark beetles invaded much of the vulnerable remaining forest, claiming another 1.5 million acres of coniferous forests. With few tree roots to grip the earth, erosion and uncontrolled run-off filled in or choked off thousands of streams. The devastation forever changed the landscape of Japan.

As desperate-sounding as the pine root oil project might appear, Dr. Thomas Jeffries, a microbiologist at the Forest Products Laboratory in Madison, thinks trees may provide the biofuel of the future. For twenty years, he's been working on ways to extract ethanol from wood. In overly simplistic terms, the challenge is to improve the ability of yeast to ferment xylose, a form of sugar found in trees. The task is more difficult than breaking down the sugars in cornstarch (today the primary product and process used to create ethanol). Based on research, Jeffries maintains there is enough excess woody biomass available in the United States to generate 60 to 100 billion gallons of ethanol annually. As a bonus, both the manufacturing process and the by-products of the process are a relatively clean, nonpolluting affair. The daily double is that the tree sugars used to create ethanol are tree sugars that for the most part are cast aside in the pulp and paper manufacturing process.[51] And the daily triple is that the woody materials that aren't used in the ethanol manufacturing process can be burned to generate steam and electricity to power the ethanol plants.

Dr. Vincent Chiang, a scientist at North Carolina State University, has focused on poplar as the ideal tree for conversion to ethanol. Poplar grows at a torrid pace and is the only tree with a known sequenced genome, meaning that scientists can more easily "pick apart" and isolate the various components to maximize certain qualities. Chiang's approach is two-pronged: (1) to maximize the process for deriving ethanol from poplar and (2) to create a version or hybrid of the tree that can be converted more easily. Chiang states, "If we can find the regulators that tell a tree to make more of one component and less of another, then we can engineer trees that are enriched with polysaccharides—a perfect feedstock for ethanol production."[52]

Dr. Gerald Tuskan, a plant geneticist at the Oak Ridge National Laboratory, is equally enthusiastic about the prospects of creating ethanol from poplar trees, which grow naturally and thrive in areas ranging from Alaska to the California Baja. He believes it will be a large-scale reality by 2017. He's discovered that poplars can be bred so the above-ground portion of the tree excels at generating cellulose for ethanol conversion, while the underground root portion favors lignin, which is better as sequestering carbon dioxide, a greenhouse gas. "It's like two different plants in one organism."[53]

In northern Minnesota, hybrid poplar plantations are already in full swing. For optimum production per acre, the trees are planted, cared for, and harvested pretty much like giant cornstalks. Treated as a cash crop, the trees can grow to 60 feet in ten years, and poplar fields can yield up to six times as much wood per acre as natural forests. Trees can be planted on marginally productive cropland, and they prevent erosion to boot.[54]

The power unleashed from the Japanese pine root experiments at the end of World War II paled in comparison to the energy unleashed over the towns of Hiroshima and Nagasaki in 1945. But just as trees had been used in an attempt to destroy, they also served as a way to restore. It's reported that trees at the centers of these bombed cities sent out new shoots from their roots within two months. John Hershey, in his book *Hiroshima,* offers this description: "Over everything—up through the wreckage of the city, in gutters, along the riverbanks, tangled among tile and tin roofing, climbing on charred tree trunks—was a blanket of fresh, vivid, lush, optimistic green."

Wood for powering planes and aircraft carriers—now doesn't that beat all? But it's no secret that wood has been in the thick of things to get us from point A to point B. It's the raw material we've used to create the floating, flying, and rolling objects that have allowed us to explore and travel—by land, air, and sea. Next are a few of those journeys.

CHAPTER 9

Wood by Land, Air, and Sea

One foggy October day in 1707, a fleet of British warships was heading back home after yet another tiff with the French. They were taking care to navigate well clear of the rocky Scilly Islands off the coast of England when suddenly those aboard the flagship felt the explosive shattering of wood. Within minutes the hull was ripped open, and the ship sank. Three ships trailing behind met the same fate. In a flash, over two thousand men drowned, and the British Navy was dealt a stunning defeat by one of their greatest adversaries of the day—the lack of an accurate way to determine longitude, and thus location, at sea.

The secret to determining longitude lay in developing a device that could keep extremely accurate time despite the temperature swings, humidity fluctuations, and violent contortions of a ship at sea. If one knew the exact time in two places—say Greenwich, England, and wherever the ship was at sea—one needed only to do a titch of spherical trigonometry to determine the position.[1] To those of us today who can purchase an accurate-to-the-second watch for $4.99 and four box tops, this may seem a trivial matter, but three hundred years ago it was a matter of life or death.

This may all seem a long way from our topic of wood—but no. In a feat that one source credits as "one of the most remarkable mechanical achievements of the eighteenth century," a self-taught carpenter's son by the name of John Harrison crafted a marine chronometer prototype that *was* accurate enough to determine longitude.[2] And as counter-intuitive as it may seem, the material that offered the highest degree of accuracy was wood. Metal could be machined to more exacting tolerances, but the lubricant of the day—animal fat—was gummy, required constant replenishing, and didn't smell that great on a hot day. So, in a stroke of blue-collar ingenuity, Harrison turned to lignum vitae; a fine-grained wood of incredibly high density that, because of its high resin content, was self lubricating.[3] He used it for the rollers, bushings, guide discs, and other parts of the mechanism subject to friction. He also used oak for the teeth and wheels of many of the gears. Wood became a hero.

Wood's role in the realm of transportation is a rich and sometimes unexpected one. In the case of the *Spruce Goose*, it played a very large role indeed.

THE SPRUCE GOOSE MADE OF BIRCH

Howard Hughes's legendary *Spruce Goose* was many things: gigantic, controversial, expensive, one of a kind. But one thing it was not: spruce—at least, not 95 percent of it. The only portions of the aircraft created from spruce were the spars used to support the wings. The primary wood used was birch. But a name like "Birch Perch" clearly lacked the poetic imagery of the settled-on nickname.

The idea for the *Spruce Goose*, originally named the HK-1 and originally nicknamed Hercules, was hatched by Henry Kaiser, who oversaw the construction of 1,490 ships during World War II. He was appalled by the fact that America had recently lost 800,000 tons' worth of supply ships to German U-boats, and he proposed a "flying cargo ship" that could cross the Atlantic safely. With $18 million in government financing—an amount to which billionaire Hughes later added $7 million more—the two teamed up in 1942 to develop and construct the plane.

Part of the government's concern was that the plane would deplete the nations already waning steel, aluminum, engineering, and manpower resources. The duo quelled these fears by vowing to use wood for the structure, engineers already on staff for the design, and non–combat age woodworkers for the construction. Furniture builders from around the country were recruited to work on the project. The woodworking firm of Weber Showcase and Fixture Company was contracted to build bulkheads for the hull. The plywood panels were built by Jasper Wood Products.[4] Louis Tribbett, one of the engineers on the project, explained, "The Hughes company just went out and hired a bunch of Swedes and good woodworking men."[5]

The *Spruce Goose* was the largest aircraft ever built in its day and is mammoth even by today's standards. The wingspan exceeded the length of a football field by 20 feet, and the tail stood eight stories tall. It tipped the scales at 400,000 pounds. The hangar erected in which to build the flying boat was just as impressive; at 750 feet in length and 250 feet in width, it was one of the largest wooden structures in its day. The plane was powered by eight 3,000-hp Pratt & Whitney engines, each with a propeller 17 feet in diameter.

Six types of wood were used: spruce, poplar, maple, balsa, birch, and basswood.[6] Hughes, the consummate perfectionist, actually sent lumber experts into the forests in Wisconsin and Canada to earmark specific trees, then had these experts follow the trees all the way to the sawmill to ensure quality control.[7]

Both the framework and the "skin" were made of wood, using a process known as Duramold. Components were made using 1/32-inch-thick layers of birch wood veneer with alternating layers running perpendicular to one another. Birch was selected because it was both lightweight and strong, it was easily molded, it held nails better than spruce, and it was more resistant to splitting, dry rot, and parasites.[8] Layers of the veneer were bonded together using plastic adhesives, and shaped using steam. It took 8 tons of nails to "clamp" the skin to the hull and wing frameworks; once the adhesive had cured, these fasteners were removed with specially designed nail pullers.[9] The resultant material was both stronger and lighter than aluminum.

In typical Howard Hughes obsessive-compulsive fashion, seventeen different versions of Duramold were developed before a final composition was decided upon.[10]

The exterior was finished with a coat of wood filler, followed by a coat of sealer. The sealer served as an adhesive for a layer of tissue paper that was applied by wallpaper hangers. This was followed by two coats of standard spar varnish and an additional coat of aluminized spar varnish. The process created a finish comparable to that of a grand piano.

Even when aluminum became available for the project in 1943, Hughes continued to forge ahead with wood. Engineer John Parkinson stated, "The story I got from my friends was that by the time aluminum became plentiful, his [Hughes's] people had gotten so sold on

The *Spruce Goose* floating on water. Despite its nickname, the *Spruce Goose* was a mere 5 percent spruce. The largest aircraft of its day, built primarily by woodworkers and paper hangers, it weighed 400,000 pounds, stood eight stories tall, and had eight propellers, each 17 feet in diameter.

this pioneering plywood design that they didn't want to change. They thought they were building a better, lower-weight airplane using the plywood techniques that they developed. But it didn't turn out that way because the stuff started coming apart."[11] Hughes himself viewed the material with a more jaundiced eye after the project drew to a close. "You might as well have built a ship that big out of putty," he bemoaned during an interview with the *Washington Daily News* in 1947.

While the *Spruce Goose* yearned to take flight during World War II, another large wooden airplane did—and did so in impressive fashion. The British Mosquito was made primarily of plywood, specifically 3/8-inch-thick Ecuadorean balsa wood, sandwiched between thin sheets of Canadian birch plywood. Casein wood glue was initially used for joining the layers, but after several unexplained crashes in tropical climates, a more durable formaldehyde-based glue was used. This plane too had its nicknames: the Wooden Wonder and the Timber Terror.

Concrete molds were used to form the left and right sides of the fuselage, which were then bolted and glued together. The wings and most other structural parts were likewise made of wood. The planes consumed minimal amounts of precious wartime metal beyond the fifty thousand brass wood screws used to hold each plane together. The wood planes had the added advantage of being able to be constructed and repaired by carpenters and furniture makers. And they were FAST—so fast that initially they were unarmed, since they could outrun even the fastest German fighter planes.

The planes were used primarily as bombers, flying twenty-eight thousand missions and dropping 35,000 tons of bombs. All in all, sixty-seven hundred Mosquitos were built and used during the war. The aircraft so incensed Hermann Goring that during one speech he stated, "It makes me furious when I see the Mosquito. I turn green and yellow with envy. The British, who can afford aluminum better than we can, knock together a beautiful wooden aircraft that every piano factory over there is building."

Even ten years after the war, Mosquitos could be found transporting cargo about, even in the brutal tropics, at 400 miles per hour.

Alas: the *Spruce Goose* eventually laid an egg—and it wasn't golden. World War II ended before the plane ever took flight. Hughes's HK-1 project eventually ran up a tab of $47 million. The Senate Committee on National Defense charged Hughes that the "flying lumber yard" was a waste of taxpayers' dollars and that the contraption would never fly. In response, Hughes returned to California to conduct low-speed taxi test runs on the water. But instead of remaining waterborne, on the third pass Hughes lowered the wing flaps, and the *Spruce Goose* flew. It only flew at an altitude of 70 feet, but it flew.

Mel Glaser, an engine mechanic who was on board for the unexpected test flight, recalled in a 1997 interview, "I didn't think he'd take it up. After all, there was no emergency gear on the plane, no life vests, and no training for a possible crash. But knowing Howard, it didn't surprise me. When he got it up to 95 miles per hour, it suddenly got quiet. I knew then the plane was airborne."[12]

The plane flew only a mile, but it was enough to silence most critics. After that short flight the plane was grounded. The eccentric Hughes continued to spend a million dollars a year for the next thirty-three years to keep it flight ready.

While using wood for airborne endeavors may seem dim-witted today, it was the material that allowed man to first take flight. Wood was the substance Leonardo da Vinci used in his doodlings in the Codex Atlanticus as early as 1480. His designs for the flapping wing, aerial screw, and flying machine all incorporated wood.

The struts, framework, propeller, and takeoff mechanism of the Wright brothers' plane were all crafted from wood: ironically, spruce. Their 1903 patent application for their flying machine spells out: "The front and rear spars of each aeroplane [wing] are connected by a series of parallel ribs which preferably extend somewhat beyond the rear spar, as shown. These spars, bows and ribs are preferably constructed of wood having the necessary strength, combined with lightness and flexibility."[13] Surely Wilbur and Orville couldn't have predicted that their flying machine, powered by an 8-horsepower engine, would give birth to the 200-ton flying boat with three thousand times the horsepower built by Howard Hughes just forty years later.

Wood stayed in the picture in the post–Wright Brothers Era. The fuselage of the 1913 Deperdussin Racer was built from three layers of tulipwood veneer applied to a frame built of hickory. The skin and framework of most World War I fighting planes, including the mighty Fokker, were made of ash, spruce, and/or plywood.

Wood is still used today. Propellers for the Shadow 200, a tactical unmanned aerial vehicle used for surveillance in the Mideast, are made of laminated sugar maple veneer. Even in this day and age, the armed forces can find no better substitute. Sensenich Wood Propellers in Plant City, Florida, continues to crank out these and other wood propellers.[14]

While the *Spruce Goose* never flew again, it did move again—twice. After Hughes's death in 1976, no one quite knew what to do with the monstrosity. At one point it looked as if the Smithsonian would lop off a 51-foot wing section, eight other museums would take smaller parts, and the rest would be scuttled. Eventually deals were struck, and the flying boat made a 6-mile journey by barge to its first resting place in Long Beach, California, where it was exhibited alongside the *Queen Mary* as a tourist attraction. In 1990 the *Goose* was moved again, this time 1,055 miles to the Evergreen International Aviation Museum in McMinnville, Oregon, where it rests today. This move entailed disassembling much of the plane, then restoration and reassembly—a task that took more than ten years.

GO FLY A PERSON: KITES FOR WORK AND PLAY

Many of us think of man's first winged flight as the one Orville Wright took when he traveled 120 feet through the air on December 17, 1903, at Kitty Hawk, North Carolina. But Sir George Cayley's coachman would take exception to that, for in 1853 he was secured to a gigantic contraption made of kites by his employer, who proceeded to fly him across a wide valley. Upon landing, the unwilling aeronaut proclaimed, "Please, Sir George, I wish to give notice. I was hired to drive, and not to fly."[15]

Kites were used for human transportation well before 1853. Hundreds of years previously, Samoans used wood-frame kites to power their canoes from island to island. Ben Franklin—best known for using a kite for electrical experiments—used one as a youth to propel himself across a pond while floating on his back. English schoolteacher George Pocock patented his *char-volant* in 1826. It consisted of a five-person carriage that could reach speeds of up to 20 miles per hour when drawn by two enormous kites. He could steer using a series of control lines, he could tack into the wind, and some of his journeys exceeded 100 miles.[16] He took enormous pleasure in the fact that his contraption was exempt from paying toll fees on certain roads, since only carriages drawn by "horses, mules, donkeys, and oxen" were earmarked as required to pay the fees. Would you expect anything less from a man who used a kite to loft his own young daughter, Martha, 300 feet into the air while she was seated in a chair?[17]

Early kites, as expected, were constructed using strong, lightweight woods. Ben Franklin used cedar for building the kites he used in his electrical experiments.[18] Kites of the Far East were invariably constructed using bamboo (not officially a wood, but a woody material). Lawrence Hargrave, who greatly advanced the design of box kites in the 1890s for use in meteorological experiments, used laminated redwood.[19] Charles Lamson lofted himself for thirty minutes to a height of 50 feet using a kite made of fir and canvas.[20] A man-carrying kite that successfully flew in the early 1900s was made entirely from broomsticks.[21] European and Australian kites were generally framed using ash, birch, and spruce, all of which are not only strong and resilient but easily bowed and shaped after being soaked in water.[22]

Though there are many accounts of people being injured and outright killed using "man-lifting" kites, it was operator error, string malfunction, and weather conditions that were invariably the cause, not structural failure of the kite framework. In 1855, Dr. Jules Laval used a kite 80 feet tall and 70 feet wide to hoist an eleven-year-old boy, seated in a wicker chair, 80 feet in the air. In a Keystone Cops–like series of events, the friction from the rope securing the kite burned the hands of the five men holding it, so they tethered the flying line around a tree stump. The continuing friction caused the stump to ignite, which in turn severed the line. The "passenger" survived a 30-foot free fall to earth. Two years later, amateur aviator Jean-Marie Le Bris experimented with a horse-drawn cart to pull a kite to which he was attached into the air. While Le Bris was airborne the horse bolted, and the flying line wrapped itself around the carriage driver, who was also lifted into the air. Amazingly, the ensuing crash resulted only in a broken leg for Le Bris.[23]

Early kites were working kites. They were used for carrying fishing lines out to sea, and early Japanese temple builders used kites to lift roofing tiles to workers on roofs. As progress marched on they continued to be used for raising telephone wires, aerial photography, and meteorology. When the first bridge was built between Canada and the United States, a kite-flying competition was held to establish a string connection between the two shores; the kite string was then used to pull a series of gradually larger ropes and cables across the Niagara River to start the project.

Perhaps kites have had the most impact in military applications. The earliest stories go way back. In 202 BC, a Chinese general and his troops were cornered. No way out. Not much in the way of supplies except for an aeolian harp, a dumb kite, and some string. Hmm. Legend has it that the general secured the harp to the kite, then, under cover of darkness, flew the kite over the enemy camp. The strong breezes played the harp strings, the opposing army surmised it was the gods issuing warnings, the enemy fled in horror, and the ingenious general and his troops escaped. Around the same time, another story describes how Chinese General Han Hsin flew a kite over the walls of a city he was about to attack, then measured the string to determine how far his troops would have to tunnel to get beyond the walls for their surprise attack.

But kites have been used more recently. In 1855, Admiral Sir Arthur Cochrane successfully used 12-foot kites to tow torpedoes to targets over 2 miles away.[24] Around 1900, S. F. Cody developed a kite that could lift a man equipped with a camera, a telescope, firearms, and a telephone for the purpose of scouting out enemy positions. He succeeded in lifting a man nearly 4,000 feet into the air—a record that arguably remains unbroken today. During both World War I and World War II, the Germans towed man-carrying observation kites behind their surfaced U-boats in order to better scout out enemy targets and position.[25]

The U.S. Navy also used kites in two unique ways during World War II. Fleets of wood-frame box kites attached to 2,000-foot-long wire lines were flown above ship convoys to shear the wings and en-

Polar explorer Roald Amundsen takes flight using a chain of man-lifting kites during experiments in 1909.

tangle the propellers of low-flying enemy planes. And the Gibson Girl survival kite was used by downed seamen to signal their position at sea, hoist a radio antenna, and provide a means of propulsion. In the not-too-distant past, large kites were banned in East Germany for fear that someone might use one to fly over the Berlin wall.[26] Another unusual use was found during the Civil War, when Union soldiers used kites to drop pamphlets explaining Lincoln's amnesty program over rebel encampments. Some six hundred years earlier, in China, kites were used to drop pamphlets among prisoners inciting them to riot.[27]

Kites come in a wide array of shapes, sizes, and capacities. The largest weighs over 2 tons and has a surface area approaching 7,000 square feet. The smallest kite to ever fly is less than 1/4 inch in height. A single kite has flown to a height of 2 1/3 miles, and a train of kites has attained an altitude of over 6 miles. The speed record for a kite is 120 miles per hour.[28]

Though kites for most of us today conjure up only images of a lazy Saturday afternoon activity, or of a bespectacled man in a powdered wig experimenting with electricity, kites played a vital role in the development of the airplane wing. Kites and kite-glider hybrids went a long way to help the understanding of lift, thrust, wing shape, and warp control. The Wright brothers used them as an early form of propulsion and experimentation on their route to developing their flying machine. The first powered aircraft were basically large box kites with motors. There are worse things to be told than to "Go fly a kite."

TRAINS: RIDING THE WOODEN RAILS

Early railroads in America may conjure up images of "riding the iron rails" and "steel horses," but those images are built from the wrong material. If you were standing beside a segment of track stretching across the United States in the early 1800s, this is what you'd see: railcars made of wood, rolling on wood wheels reinforced by a thin strap of iron, riding on wood rails (yes, rails), supported by wood railroad ties. The fireman would be pitching chunks of wood into the firebox. And in many cases the cargo being hauled would be lumber. Wood was, and still is, a major role player when it comes to trains.

The earliest railways were actually wagonways built in England as early as 1604, primarily for hauling coal. The tracks were constructed of parallel lengths of oak, ash, and birch, supported by ties every 2 to 3 feet, which in turn supported carts with specially designed wood wheels to ride the rails. The engine was an example of true horsepower, since locomotion was provided by horses.[29]

Early railways were similar in America. Since wood was so plentiful and iron so expensive, early tracks were constructed in the manner described by one early Michigan settler: "They took timbers as long as trees . . . hewed them on each side and flattened them down to about a foot in thickness, then laid them on blocks which were placed in the bed of the road. They were laid lengthwise . . . far enough apart so that they would be directly under the wheels of the cars." These tracks were usually topped by straps of iron to increase longevity.[30]

The abundance of wood in the New World made American and English railways considerably different in two respects. First, constructing a mile of American track cost about one-sixth that of a track in timber-poor England. Second, while English trains primarily burned coal, early American trains—up through the Civil War—burned wood.

Since early steam locomotives consumed so much wood and water, "wood-up" stations were built every 10 to 25 miles along the tracks. One wood yard at Columbus, Nebraska, measured half a mile in length and stocked a thousand cords of wood. In the 1850s, steam engines devoured four million to five million cords annually. An entire cottage industry supplying the railroads with firewood and water grew up along the tracks.[31]

Wood-burning trains had their flaws, the most notable being their propensity to catch things on fire—including the countryside they rode through, the train cars they pulled, and the passengers within. Charles Dickens referred to the sparks put out as "a storm of fiery snow," but the results were anything but poetic. The railroads were accused of burning more wood as they passed through wooded areas than they consumed in the firebox.[32]

Early wooden railcars were often roofless, making a train ride a treacherous form of transportation. Passengers would bring umbrellas to protect themselves from the sparks that were spewed, but once the umbrella covers had burned through, things became interesting. Women's dresses, usually made of flammable material, often ignited. One account claims that women "were almost denuded" by the sparks.[33] Some railways issued buckets of sand to passengers so they could douse one another if they caught on fire. In another incident, $60,000 worth of newly minted dollar bills burst into flames when sparks landed on them.[34] Taller smokestacks solved some of the problems, but some were so tall that they needed to be hinged so they would fit under trussed bridges. Early wood-burning trains posed such a fire hazard that railroad employees followed trains across wooden bridges with buckets of water to douse any flames.

But it was the fires started in the fields and forests along the tracks that wreaked the most devastation. Despite over a thousand patents issued for spark arresters and other safety devices, fires along the tracks were an everyday occurrence. Slash and other branches left along the tracks by people clearing land and cutting wood for the engines turned certain areas into tinder boxes.[35]

The Hinckley fire, which in the fall of 1894 took 418 lives, was started by a spark from a train. It was a super-heated firestorm, which burned so intensely that buildings burst into flames before the fire reached town. Smoke could be seen 80 miles away in Duluth. Big-city folks weren't immune to the dangers of early wood-burning locomotives, either; even in Manhattan, people on the streets below bridges were burned by the cinders of locomotives passing overhead.[36]

Despite the hazards, many refer to this time as the golden age of railroads. Wood-burning locomotives burned fuel relatively cleanly, unlike the sooty, coal-burning models that followed. The 1860s ushered in the ornate Victorian era, and wood-burning trains followed suit. Locomotives were workhorses, showpieces, and public relations departments all rolled into one massively impressive package.[37] When the new-fangled trains rolled into town, it was a major event.

These early wood burners are the ones characterized in "the little train that could" children's books—colorful, cheerful affairs with four huge drive wheels, a large funnel for a smokestack, and glittering chrome. Engineers were equally glittery and impressive. One source explains:

> An engine driver at the beginning of the railway era was a much-respected technician of no mean social status—something akin to an airline pilot of the present day. He controlled a powerful, complex, and somehow mysterious machine. He was the well-paid master of one of the most glamorous mechanisms on the face of the earth. Wood burners were clean: the light ash blew away, and the upkeep of their ornamental surfaces was a relatively simple duty. The engineer dressed the part of his august station, and from surviving pictures we see him as something of a dandy among workingmen. But the coal burners changed all that. An oily grime dusted both man and machine with a stubborn black deposit. Fanciful decoration gave way to plain black paint; suits and vests gave way to practical dungarees and overalls. The locomotive became more common and workaday. Its master became just another segment of the labor force who at the end of the workday was almost indistinguishable from a factory hand.[38]

The trestles built to carry tracks and trains across deep gorges and wide valleys consumed monstrous amounts of wood. The Red Sucker trestle—part of the Canadian Pacific Railway's "200 miles of engineering impossibilities" built in the 1880s—was 110 feet high and 900 feet long. The Mountain Creek Bridge, part of the same project, was built using 1.5 million board feet of lumber.

Even today, trains run on wood. The seven hundred million railroad ties that support the 220,000 miles of track in the United States are replaced at a rate of fifteen million to twenty million ties a year.[39] Though concrete and composite substitutes exist, 95 percent

The Red Sucker railroad trestle, part of Canadian Pacific Railway's "200 miles of engineering impossibilities," was 110 feet high and 900 feet long.

of all ties in the United States remain wood; they have the necessary give, spike-holding ability, and, when treated, longevity, which makes them difficult to beat. John Fristoe, the first president of the Railway Tie Association, referring to trees sawn into railroad ties, philosophized in 1924, "[a] tree could plan no greater destiny than to give up its existence, not in old age but in its prime for the service of mankind."[40]

IN SEARCH OF THE LOST ARK

One of Bill Cosby's early comedy routines was based on a mythical conversation between Noah and God. God tells Noah he wants him to gather wood and build an ark that's 300 cubits by 80 cubits by 40 cubits.

"Right," replies Noah. And then, after one of Cosby's trademark pauses, he asks, "What's a cubit?" It turns out even God doesn't know.

As the routine progresses, one of Noah's neighbors, late for work, says "You wanna get it out of my driveway? I got to get to work. Listen, what's this thing for anyway?" When Noah tells him it's a secret, and the neighbor presses him for a hint, Noah responds "Well, how long can you tread water?"

It's a hilarious routine. And the notion of building an ark the size of an aircraft carrier out of wood, loading it with pairs of animals, and weathering forty days and forty nights of torrential floods is an equally hilarious notion to many. But not to everyone; especially the Turkish government, which in 1987 built a visitors' center near Mount Ararat; the site of what many believe to be the final resting spot of the biblical Noah's ark. So what evidence is there to back up the ark site? Ron Wyatt, a biblical archaeologist with a particularly strong, and undeniably biased, Christian bent, offers the following:

> The outline found on the site is in the shape of a boat and is the same length mentioned in the biblical quotations—300 cubits. An Egyptian cubit, at about 20.6 inches—the distance between the elbow and the tip of middle finger—puts the footprint at 515 feet. For the record, Bill Cosby was wrong in two of his measurements: the Lord designated a width of 50 cubits and a height of 30 cubits.[41]
>
> It's located in eastern Turkey, as hinted in the Bible.
>
> Petrified wood, manmade metal connectors, and anchor stones have been found in the area.
>
> The site was a tourist attraction long, long, long ago. The historian Flavius Josephus who lived in the first century, mentions this in one of his early works.
>
> Radar scans show a regular pattern of timbers, bulkheads, keels, and so on.

One must bear in mind that the above was written by a man who claims to have discovered sulfur balls from the burning of Sodom and Gomorrah and of whom one member of the Israel Antiques Authority stated, "[his claims] fall into the category of trash which one finds in tabloids such as the *National Enquirer*."[42]

But Wyatt is surely not alone in his belief that remnants of the ark exist. Marco Polo, who traversed the area in the 1200s, made note of the legend, indicating that remnants of the ark could be seen in those early days. In 1916, Czar Nicholas of Russia sent two expeditions to Mount Ararat to photograph an alleged ark. In 1949, American reconnaissance planes photographed the area, and rumors flew that the photos contained images of the ark. The photos remained secret until the Freedom of Information Act in the 1990s but, when finally shown, revealed nothing conclusive. In 1977, President Jimmy Carter is rumored to have had his plane fly over Mount Ararat to take a look for himself, on his way to visit the Shah of Iran.

Some believe in the ark but believe Turkey is the wrong spot. In 2006, a dozen people, including archaeologists, geologists, biblical historians, and a group of Texas businessmen, went searching for the ark in Iran. At 13,000 feet they came across a 400-foot man-made object, consisting in part of petrified wood resembling a ship. A Houston laboratory also found fossilized sea animals buried inside. Arch Bonnema, a member of the expedition, flat-out asks, "How did a ship get to 13,300 feet, except to float there?"[43]

Regardless of all the ifs, whens, and wheres, the ark described in the Bible was a BIG project. For it to have accommodated all the animals, food, and water, calculations put the volume at over 1.5 million cubic feet; the equivalent of 569 standard boxcars or enough room for 125,000 sheep. But it was also a LENGTHY construction project. Noah was given the work order when he was 480 years old, and he worked on it for 120 years until the rains started coming (and coming and coming).

Since the ark really didn't require steering—where was it going to go?—the shape was most likely boxy rather than streamlined. The biblical version was three stories high. Based on modern calculations,

This ark replica, built by Johan Huibers of the Netherlands out of cedar and pine, is half the length of Noah's. It's inhabited by life-size model elephants, giraffes, and crocodiles, and is open to the public for tours.

the designated dimensions would have made it extremely stable, capable of righting itself, even when tilted to nearly 90 degrees. What it might have been made of is the subject of much conjecture.

In Genesis 6:14, God tells Noah to "Make yourself an ark of cypress wood, make rooms in the ark, and cover it inside and out with pitch." But other interpretations maintain that Noah was told to use "gopher wood." And what that is, is open to much debate. Some folks maintain that a scribe incorrectly transcribed the word "kopher," which means a wood that's covered in pitch. Others maintain that "gopher wood" was a wood of massive strength that no longer exists today, perhaps having vanished in the big flood itself. Various translators have interpreted the wood to be pine, cedar, fir, ebony, juniper, boxwood, and slimed bulrushes.[44] Still others interpret "gopher wood" as meaning squared beams; others as "laminated beams."

Of course, for most people the quandary isn't over what type of wood the ark was crafted from but whether it was crafted at all. Many cultures have a similar type of doomsday flood tale in which a deluge kills all but a chosen or lucky few. There's the Babylonian epic of Gilgamesh, the Greek and Roman saga of Deucalion and Pyhrra, and even a Native American legend similar to that of Noah's ark.

Jose Solis, a doctoral candidate in the College of Architecture at Georgia Institute of Technology, designed and built a model of Noah's ark based on scientific calculations. He determined, based on the available woods of the day, that the ark itself would weigh 3,676 tons. Animals, food, and storage would add 4,560 tons, and water and a water storage system would add another 4,000 tons. He determined that the ark would have needed to be 99,150 square feet to accommodate everyone and everything, and based on current lumberyard prices, the cost would be $16,472,040. His model was a three-story affair, constructed using trusses, that looked much more like a barge than the children's storybook image most of us conjure up. But it's actually a very attractive-looking barge.[45]

Doubts will always abound; yet, enough people are curious about the matter to have inspired several travel agencies to now offer tour packages that include working expeditions to the area where the ark is believed to have landed. And it's no easy trek. It's in a politically unstable area, and part of the region is within a military zone. The peak, where many believe the ark resides, is 3 miles above sea level and covered by a 17-square-mile ice cap hundreds of feet deep in places. The terrain is rugged, and the season that vaguely resembles summer lasts only six to eight weeks. The locals call it Agri Dagi, which is Turkish for "mountain of pain."

THE SONG OF THE GONDOLIER

Even with the hustle, bustle, tourists, and noise, it *is* romantic. My wife, Kat, and I recline on the tapestry-clad seat, listening to the sweet strains of accordion music. Guilermo, our gondolier, rows the craft silently and effortlessly. As we look ahead we can see the classic asymmetry of the hull, the ornate bow decoration, and the "Gas: $2.49/gal" sign. Huh?

We may be sitting in a Venetian-built gondola, but we're 4,700 miles from the City of Water. We're on the St. Croix River, which divides Minnesota and Wisconsin; the music is coming from a boom box behind us; and Guilermo, when he takes off his straw hat and striped shirt, will be just plain old John Kerschbaum.

Theoretically, Kerschbaum could row us all the way to the squero, or boatyard, where his gondola was built. It would involve following the St. Croix and Mississippi Rivers to the Gulf of Mexico, crossing the Atlantic, wending through the Strait of Gibraltar, then working his way through the Mediterranean and Adriatic Seas. But for now his goal is to make it the next 50 feet without being bisected by one of the 300-horsepower cigarette boats racing toward us.

The canals of Venice are just as crowded as our current setting and narrower to boot—some a scant 12 feet across—and that is where the whole story starts. To create boats that could pass one another in the Venetian canals and make tight 90-degree turns, the design evolved so only a single oar was needed. A gondola is a very specific thing. It is 36 feet 2 inches long and 4 feet 10 inches wide, and it weighs 900 pounds, give or take a pound or two. It is flat bottomed, so it can navigate in water as shallow as 1 foot while carrying a cargo of 2,000 pounds. It contains 280 pieces of wood—some decorative, most structural, almost all different because of the asymmetry.[46]

It is ingeniously lopsided. The left, or port, side is 10 inches (24 centimeters) longer than the right, or starboard, side, which makes the left side rounder and the right side slightly flatter.[47] This length difference also tilts the entire boat about 9 degrees to the right.[48] Gondolas are rowed from the right to counteract this imbalance. The natural tendency of the gondola to veer to the right, because of its shape, is counteracted by the push of the oar, which tends to push the bow more to the left. The boats have a tremendous degree of rocker, or curvature from end to end, so only half the bottom is below the water line; this makes the boat easier to maneuver and turn in the tight canals. "Everything in the boat has a purpose to it," explains Kerschbaum. "But when you work on a boat for a thousand years, you get it right."[49]

A gondola is a floating work of efficiency and ergonomic art. "It's no harder to row with four people in it than it is with one," explains Kerschbaum. "With more weight, you get more momentum, and it just slices through the water. Once you get the thing going, it takes the same amount of energy to row four people as it does for one person to walk."

Eight different woods, each with its own special qualities and strengths, are used to build the traditional gondola. Oak, solid and strong, is used for the planking on the sides. Lightweight fir is used for the bottom. Elm, sturdy and tough, is used for the frame. Cherry, easy to shape, is used for the thwarts. Mahogany, lightweight and water resistant, is often used for the decking. Walnut, dense and beautiful, is used for the forcola, or oarpost. Linden is used for the sternpost. Larch is used to reinforce the tips of the bow and stern.[50]

For a short period, some gondolas built in the mid-twentieth century were constructed partially of plywood. But no more. The Municipality of Venice passed rules mandating that only solid wood could be used for public service gondolas. Proponents of plywood maintain that their boats have longer life and require less maintenance. Traditionalists scoff. The El Felze association, a consortium of those building traditional gondolas and components, commented, "It was a right choice to protect the gondola's tradition. The gondola is a world-famous symbol of Venice. It cannot be modified. If we don't define rules, in the future we risk seeing gondolas made of plastic."[51]

There may be an appropriate use for plastics in the canals of Venice. There are over 200,000 *paline* and *briccole*—a forest of vertical posts that emerge from the canals to serve as navigational aids and tie-up spots for gondolas. This "urban Venetian furniture" comes in many guises. Rows of single posts delineate pathways for boats; a group of five posts encircling a single taller one designates an intersection of canals; piers decorated with stripes or spirals and topped by decorative "hats" mark out turf in front of houses of historical families and institutions. Unfortunately, water-based parasites find these posts as alluring as the citizens and tourists of Venice. A post's average life span is seven years; five years for those in major traffic areas. Each year, thirty thousand posts are replaced—an activity that taxes the area's forests and economic reserves. Some are pushing for the undeniably less aesthetically pleasing, but undeniably more functional, plastic or composite substitute.[52] But for now, tradition rules.

Gondolas, in one form or another, have a tradition dating back a thousand years. They were first mentioned in writings in the eleventh century as being used for commerce and transportation for those living on the 118 assorted islands that constitute Venice. By the fourteenth and fifteenth centuries, gondolas began appearing in Venetian art. Their popularity peaked in the seventeenth century, when over five thousand gondolas plied the waterways of Venice. At one point, enclosed cabins, or felzes, which offered protection from the elements and privacy for conversation and romantic interludes, were popular.

In 1797, the Republic fell to Napoleon, and the gondola's fortunes fell with it. Economic necessity prompted boatbuilders to experiment with designs that would allow gondolas—which up to that point had usually been manned by two gondoliers—to be rowed by a single oarsman. In the late 1800s, a boatbuilder by the name of Domenico Tramontin perfected the ingenious asymmetrical hull used now by the gondolas gliding through Venice.

In an ironic twist of fate, it is the modern tourist who keeps the Old World gondola alive. The real workhorses of the canals are now the vaporettos, or motorboats. But that doesn't mean a gondolier can't make a decent living. "It's a great job," says Kerschbaum. "They make a *lot* of money. And a lot of it is cash under the table." The city of Venice issues only 450 gondolier licenses, which—like large city taxi medallions—are so hard to get they're passed down from generation to generation.

Gondoliering has always been a lucrative profession. Centuries ago, wealthy families had their own gondolas where everything from business deals to adultery took place beneath the covered felzes.

A gondola has the most distinct and recognizable shape of any boat on earth. Even in silhouette, a gondola, and Venice, the city in which most gondolas dwell, can be named by most people.

"Gondoliers were elevated from working class to upper class because they were paid well to keep the family secrets," explains Kerschbaum. "Even now there's a code of ethics: never to repeat what you've heard in a gondola."

Gondolas spread to other parts of the globe long ago. In 1661, two gondolas were given to King Charles II of England; he took a great liking to being rowed down the Thames in his unique craft.[53] The Venetian ambassador bragged, "This week [the king] was given by the states of Holland a small vessel of great beauty to sail on the Thames, but he is more gratified by the gondolas, and nothing makes him happier than to go out on the waters."[54] In 1673, two gondolas were dropped into the water at Versailles. In time, gondolas could be found in Hamburg, Amsterdam, Buenos Aires, and other cities. Gondolas have been a tremendous ambassador for Venice and Italy.

You can visit the Tramontin squero in Venice today and watch Domenico's grandson, Nedis, and great-grandson, Roberto, build gondolas using the same designs, tools, and processes as their predecessors. The process begins by waiting. Dense rot-resistant heartwood is stacked with spacers in between in the open air and allowed to dry naturally. For every centimeter in thickness, the wood is allowed a year to season. The thirty-some oak and elm frames or ribs, each varying slightly from the others, are constructed by the squeraroli (gondola builders) based on old patterns. Each rib is slightly asymmetrical to create the slightly curvier port side and slightly flatter starboard. Three of these frames—the stern, main, and bow frames—are designated as "master" frames and are spaced several feet apart in a decades-old jig along with the stern and sternpost as a starting point in the building process. Axes, hand planes, and bow saws are used for much of the work.

Two 37-foot-long oak planks, 5/8 inch thick, are precurved by applying heat to one edge and water to the other. These two long side pieces are then clamped to the three master frames. At this point, the rough shape of a gondola has been established. The remaining ribs are inserted, and more boards are installed to complete the sides. Arched crosspieces are added to support the wood platforms, or decks, that will eventually cover the bow area and provide the platform at the stern area for the gondolier to stand on. Mahogany boards are used to cover the front and back decks. Now the boat looks and feels like a bottomless gondola—and the bottomlessness is for a reason.

The gondola is turned over. And then, as if skewing the boat to the starboard side weren't unusual enough, another unique step takes place. The weight of the gondolier is taken into consideration, and the gondola is customized for that particular gondolier. Flaming bundles of marsh reeds are brushed along the sideboards to "relax" them, then wood spikes are used to press—literally, squish—the boat until it is balanced exactly right. If the gondolier is heavy, the stern is raised slightly. If the gondolier is of normal weight, the bow is raised. Only after this process are the bottom boards installed. The bottom is coated with tar pitch. The top and sides are painted black. By now, the squeraroli have spent one month crafting a black gondola that will float but has no trappings. Many would argue that the trappings make the gondola, and the makers of the trappings are just as specialized and important as the squeraroli.

As the El Felze association explains, "The gondola itself is the product of [a] complex system, the results of contributions from very different trades: it is the boat, the oar, the forcola, the brass horses, the felze, the ferro, the cushions, the carved and gilded decorations, the seating for the passengers, the hat and the clothes of the gondolier. Everything ensures that in the gondola you proceed quietly, relaxed, safe, whether alone or accompanied, and you can sing, laugh, joke, play, and do whatever you wish."[55] Who are these specialists?

Remeri are carpenters who specialize in making oars and forcola—the part many of us non-Venetians would call the oarlock. An oar begins life as a 14-foot-long plank of rough wood, usually beech, and emerges as a well-engineered hydrodynamic rowing device. Using only a plane and a few other tools, the remer creates a conically shaped grip that widens into a cylinder, then flares into a blade, considerably narrower than that of a canoe paddle.

The forcola begins with a hunk of walnut, cherry, pear, maple, or other dense wood that has been split into quarters and left to dry for two to three years. Using bow saws, bandsaws, and the remer's axe, the remeri rough-shape the forcole. As with the balancing of the gondola, the forcole size and shape is crafted based on the height and other qualities of the gondolier who has ordered the boat. Drawknives, files, and scrapers coax the log into its final shape, and in the end you have a vertical post with a variety of crooks, cutouts, and bends upon which the gondolier can rest the oar in order to maneuver his craft. The completed forcole has all the curvaceous, sensuous, and artistic qualities of a 1960s Henry Moore sculpture; indeed, the Museum of Modern Art in New York, Frank Gehry, and I. M. Pei have all purchased them to display as works of art. One Venetian-born writer, Sebastiano Giorgi, poetically explains: "Curvaceous and elegant, simple and complex, the forcola represents the contradictory greatness of this city. It is at the same time curious, beautiful, and functional."[56]

The fravi are the blacksmiths who make the ferro, the metal bow piece that is part decorative, part protective, part counterbalance to the weight of the gondolier on the stern. Appearing somewhat like a medieval hatchet, every part means something: Its S shape represents the grand canal; the six vertical strips, Venice's six quarters; the rounded top, the shape of the Doge's hat.

The intagiadori carve ornate or figurative designs; the indoradori gild the decorative parts, the fonderie cast the metal seahorses and other elements that adorn the boat, the tapessieri upholster the cushions, the sartori tailor the special clothes of the gondolier, the barateri make the straw summer hats and wool winter hats, the caleghieri make the shoes.[57] When all is said and done, a new gondola costs between $35,000 and $45,000.

It is only after the paint has dried and the boat is in the water that one can witness the dynamic interplay between gondola, gondolier, oar, canal, and tradition. The gondolier can execute a wide variety of strokes based on the position of the oar in the forcola. The uppermost cutout facing the stern is for going fast; the large crook below it for going backward; the crook at the bottom for starting; other designated positions are for turning and stopping. One Venetian nautical expert explains: "Every curve, corner, and protrusion has its own feeling and is the fruit of a long practical journey, that through the centuries, transformed the rough forcola of the lagoon oarsman into that of a mature object which we can admire today."[58]

I ask Kerschbaum about the lure of the gondola, whether in Venice, Italy, or Stillwater, Minnesota. "It's all about romance and nostalgia," he explains. "The gondola is the most recognizable boat in the world. Show a profile of one to people, and they'll know exactly what it is and where it came from. These boats capture the beauty and romance of Venice. People like the feeling of being cradled. They like the rocking motion of the gondolier rowing. I tell people 'With every oar stroke you get a thousand years of history.'"

While learning about those thousand years of gondola history, you'll learn about another vital connection between Venice and the world of wood: the city itself bases its existence—literally—on wood. You'll also find wood in lots of other unusual places: in outer space, as part of our urban water systems, and in the shape of the world's largest yo-yo.

Next are a few of those stories.

CHAPTER 10

Wood in Unusual Uses and Peculiar Places

When scientists at the Chinese Space Agency were looking for a material from which to construct heat shields for their single-use reentry vehicles, they turned to wood. Oak, they discovered, was a supreme ablative, self-sacrificing material; at 1,500 degrees C the 6-inch-thick oak shield would turn to charcoal, then slowly burn and surrender itself as it dissipated heat while tearing through the atmosphere.[1]

When the U.S. Navy was looking for a material from which to construct propeller shaft bearings in the twentieth century, they too turned to wood. Lignum vitae—a wood of monumental hardness for durability and high resin content for lubrication—was used on submarines and battleships alike. The *New Jersey* and the *Missouri* are just two of the better-known battleships that employed this material.

Whether it's 100,000 feet above the surface of the earth or 1,000 feet below, you can find wood. And there are lots of other peculiar places you can come across it, and lots of unusual uses you can find for it.

VENICE: THE CITY PERCHED ON WOOD

Sight down any of the 150 canals that warp and weft their way through Venice and you'll discover a city that appears nearly wood-less. Yes, you'll see black-painted wooden gondolas and rows of NBA-height wooden doors, but seemingly everything else is brick, stone, stucco, cement, and water. Which is odd for a city that bases its existence—in a very literal sense—on wood.

Venice consists of 118 small islands that were at one time very separate, very marshy, and very nearly uninhabitable. But the area had three things going for it: location, location, location. Originally it was home primarily to those seeking refuge from the Visigoths, Huns, and other barbarians wont to invade the area in the fifth and sixth centuries. But eventually its strategic location—ideally situated along vital trade routes linking Europe and the Near East—made it an area ripe for growth. The problem, however, was that there was no place *to* grow. The land was swampy. Tides and high seas regularly submerged all or parts of the islands. It was hard getting around. But then, just as now, these factors weren't enough to stand in the way of shrewd land developers. And these were developers who developed the land in the truest sense of the word.

I sit with Sebastiano Giorgi and Elena Barinova in their rooftop dwelling overlooking Venice while they clarify the "solidification" of Venice. They look like they've just stepped out of a Dolce & Gabbona ad, but there are few couples whose combined knowledge can shed better light on the subject. Sebastiano was born and raised in Venice and, though he has a law degree, edits and writes about wood things in water—boats, piers, cities. Elena, born in Siberia and educated in Paris, is an archaeologist working on excavations at Lazzaretto Vecchio, an island off Venice that was originally used to quarantine visitors during the Black Plague.

They help explain the system that evolved for solidifying and enlarging the soggy islands. Initially, fences of wattle and daub (sticks and mud) were partially submerged to create perimeters around the low-lying islands. The swampy areas around the islands were dredged, and the earth was placed within the perimeter walls and compacted to create more stable land. Eventually, both the canals that were created by dredging and the islands created from the "dregs" became better defined.

At first, the dwellings that were built on these man-made islands consisted primarily of lightweight huts built of wood. But as Venice grew wealthier and trade increased, Istrian stone, brick, and other building materials were imported to build more permanent and more lavish structures: *heavy* lavish structures, which would sink unless some type of support system was created. And the support system that was created consisted of wood pilings—millions and millions and millions of them. Giorgi uses a metaphor to explain things: "The palo [pier] is a principal foundation and can be seen as a 'hypen,' a provisional border between earth and water."[2]

A piling is nothing more than a log driven into the ground, used to support something extraordinarily heavy. Given this scenario, one would surely think Mr. Blackwell would put pilings made of wood at the very top of his list of the Ten Worst-Dressed Engineering Concepts. But let's take a closer look.

First, when a piling is pounded into the ground, the soil it displaces becomes compacted along the edge of the hole. In the same way that the displaced wood fibers of a board press against a nail shank to hold it in place, so does the dirt around a piling. And the compacted soil below the piling becomes a sort of minifooting in and of itself.

Second, wood pilings are tapered and are driven into the ground "upside down," or narrow end first. This sets up a dynamic whereby the entire surface area of the piling is "held up" by the immense amount of friction applied by the surrounding earth. And isn't it convenient that trees are naturally tapered?

Third, wood pilings have built-in engineering. When the pounding and banging will no longer drive the piling any deeper, it's telling you, "Time to stop! If I can now support the weight of a bashing 2-ton pile driver, I can surely support the weight of a static 2 tons' worth of structure." And, of course, the beauty of wood is that when the piling has reached that point, one simply lops the top off at the appropriate height.

Fourth, wood is tremendously strong in compression along the grain. If prevented from bending—as pilings are by the surrounding earth—they are nearly infallible.[3] In tests, a 15-foot-long Sitka spruce log with an average diameter of 12 inches will withstand a load of 75 tons before failing.

Fifth, as we have seen earlier in this book, when wood is totally and consistently submerged in water or some other oxygen-deprived environment, it can last hundreds, thousands, even tens of thousands of years. If the wood has natural rot-resistant properties or is treated, all the better. Wood deterioration and decay arise not in situations where wood is in constant contact with water but rather in situations where wood is alternately in contact with air, then water, then air, then water. In situations where pilings need to be extended above their watery sanctuary, they're best topped with sections of concrete at and above the waterline to create a piling that incorporates the best of both worlds.

And last but not least, timber pilings can actually withstand the ravages of acidic and alkaline soils better than those made of concrete or iron.[4] Furthermore, in situations where you want a material with a little "give"—such as docks and shipyards where large vessels dispense megaton love taps—wood pilings have the ability to absorb impact.

Pilings are not without their enemies. Termites can have a field day with untreated wood pilings in damp soil, especially in warmer climates. The heartwood of Southern pine and redwood offers some resistance, but all in all, with termites in the neighborhood, it is a losing battle.

Teredo navalis, the dreaded shipworm described by one source as "a translucent snake with vents and siphons at its tail and a clam-like head at the other end that chomps like a pair of wind-up teeth," can wreak havoc on marine wood pilings and wood boats alike. Shipworm is estimated to cause $4 billion in damages a year to wood marine structures in the United States alone. It can attack with such ferocity and stealth that ships have been found adrift at sea still tied up to sections of the "now-pilingless" wharfs to which they were secured.[5] But in the absence of marine borers, termites, and such, submerged wood—as Venice will attest—has stupendous staying power.

The slowly evolving city of Venice had little to offer in the way of the raw materials needed to shore itself up. When nearby forests were depleted during the piling process, the Venetians began ranging farther and farther. Much of the wood was cut in present-day Croatia and Slovenia. Venice's appetite for pilings was so voracious that some forests of Slovenia have yet to recover; yet the Gorski kotar region recovered to the point where it is now referred to as "the green lungs of Croatia."[6]

Excavations in Venice have unearthed a wide variety of trees used for pilings and underpinnings: Oak, larch, silver fir, elm, alder, poplar, birch, hornbeam, pine, and walnut were all used in varying capacities and amounts. In her excavations, Elena has come across well-preserved pilings 45 feet in length.

Larch, for one, makes for a tremendously stable piling. The mightiest specimens live over seven hundred years, attain heights of 130 feet, and grow as straight as the proverbial arrow. Its heartwood is naturally rot resistant and is used even today for boat planking, railroad ties, bridge construction, and other applications where wood might be subjected to moisture. It combines the best traits of others in its conifer family: it has the rot resistance of redwood, the strength of Douglas fir, and the gloriously prolific nature of white pine.[7]

Alder makes an equally fine piling for the swampy lagoons of Venice, since swampy lagoons are its home turf. So immune is alder to the ravages of moisture that it was commonly bored out and used as underground water pipe.

A variety of increasingly sophisticated and efficient methods for driving piles were developed. The earliest system involved two people simply bashing them in by hand with a heavy log resembling a conga drum with handles on either side. This evolved into a tool, roughly resembling a bladeless guillotine, by which a heavy weight, raised, then dropped by a system of ropes and pulleys mounted on a sturdy wood framework, was used for pounding. Eventually physics was harnessed, and a wheel-and-gear system was developed for lifting and dropping the pile driver.

The system worked tremendously well. In many cases, pilings were driven in as tightly packed as possible to firm up large tracts of land. In other cases, they were installed in a strategic pattern to accommodate specific structures. Each leg of the arched Rialto Bridge rests on 6,000 alder tree pilings. The chapel of Santa Maria of Health, built in 1630, rests on 180,000 wood pilings. This seems like an enormous number until one compares it with the underpinnings used to support St. Mark's Cathedral. One million pilings were driven in place, then criss-crossed by thick, massive planks that were in turn covered with earth. Upon this base the absolutely massive cathedral was built in 828 to accommodate and celebrate the arrival of Venice's patron

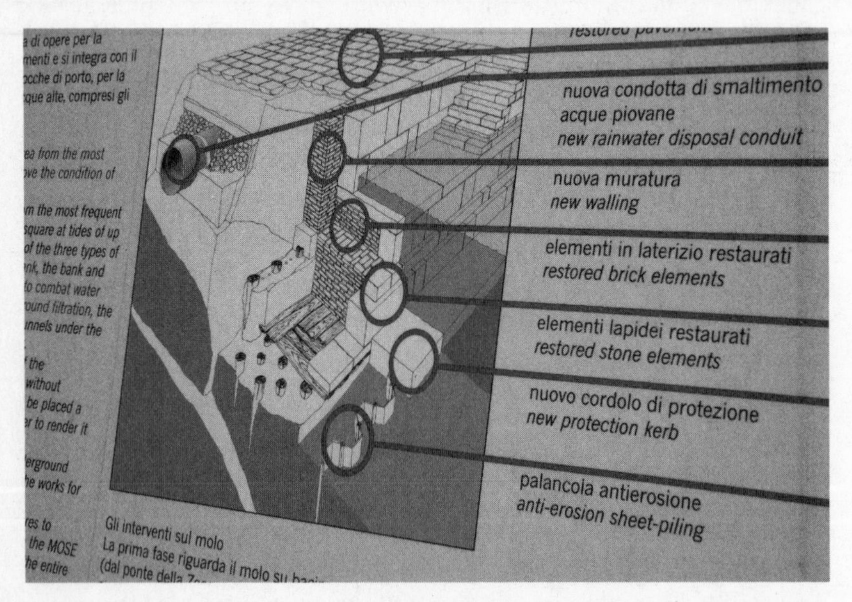

Close-up of a sign in San Marco Square in Venice explaining the process being used to protect parts of the city from further erosion and degradation. The lower portion of the diagram shows the original wood pilings and planking system used to support the structures and perimeter walls.

saint—St. Mark. (The fact that St. Mark had been dead for 750 years and that his body was stolen from Alexandria by Venetian sailors—who apparently took the order to "steal St. Mark's body" literally, because they left the head in Alexandria—seemed not to matter.)

Wood samples taken from the pilings supporting the Malibran theatre and other historic buildings in Venice were recently analyzed using radiocarbon dating and a form of tree ring analysis termed, for those who like a mouthful, "dendrochronological wiggle-matching." The still-hardworking oak pilings were discovered to date back to between AD 658 and 687. In some circumstances it was discovered that the larch pilings had become petrified by the brackish waters of the lagoon. One report explains "that the wooden piling foundations of Venice can consequently be as hard as the rock of Manhattan island. During the restoration of the eighteenth-century Jesuit church in the city's northern quarter, the steel bits of the modern drilling equipment broke on the ancient wooden pilings of the building's foundations."[8]

In 1902, engineers were given an opportunity to examine another early form of piling foundation when the bell tower of San Marco, built around AD 900, unceremoniously collapsed into a pile that could best be described—upon viewing pictures of the fallen landmark—as an anthill of enormous proportions. What they found was a relatively simple support system consisting of a few rows of short larch pilings driven around the perimeter, criss-crossed by two thick layers of planking. The bell tower had essentially been "floating" on a raft of wood with thickened edges for a thousand years.

The Venetians were not the first people to hit upon the idea of pilings. In 1620 BC, the Romans used timber piles to support their first bridge across the Tiber River. They also used wood pilings to build the first bridge across the Thames River in London in AD 60.[9] Near Rochester, England, two-thousand-year-old pilings used to support early roads have been excavated.[10] And timber pilings used as the foundation for the Circus in Arles, France, built in AD 148, can be seen in the museum on the site today. Marcus Vitruvius Pollio, born around 80 BC and considered by many to be the father of Roman architecture, wrote about the phenomenon in his *Ten Books of Architecture,* saying, "piles driven close together beneath the foundations take in the water that their own consistency lacks and remain imperishable forever, supporting structures of enormous weight and keeping them from decay."[11]

One shouldn't think of wood pilings as a thing only of the distant past. The foundation of the Empire State Building rests on wood piles,[12] as does the Brooklyn Bridge. The Louisiana Superdome and the cargo terminal at JFK Airport in New York are likewise perched on wooden piles. When a new shopping complex was built at the Tobacco Dock in London, the engineers decided to reuse the 160-year-old Scots pine piles already in place. The U.S. Army Corps of Engineers has used over six million timber pilings in constructing the locks and dams of the inland waterway system. Beginning around 1880, many pilings were pressure impregnated with coal-tar creosote to further lengthen their life span. Today, over half a million wood piles continue to be pounded in annually throughout the United States.[13]

The Roman architect Vitruvius may have been correct in his assumption about wood pilings being "imperishable forever," but that applies only to pilings situated in ideal conditions. And in recent years, Venice has been experiencing a bout of less-than-ideal conditions—it is sinking and rotting. The Queen of the Adriatic is no stranger to flooding; indeed, as you walk the streets you will see what appear to be stacks of short-legged tables. These short-legged tables are placed end to end to create elevated walkways when the ever-so-frequent moderate flooding occurs; sixty times per year is average. But the flood of 1966 was anything but moderate or average; San Marco Plaza and most of Venice were submerged beneath 3 to 6 feet of water.

Between the increasing frequency of *acqua alta,* or high water, and the decrease in elevation—the city is sinking about 3/8 inch per year—Venice's fragile equilibrium is in peril.

There are no easy or single solutions to saving a sinking city composed of thirty thousand buildings, sixty thousand residents, and fifteen million tourists. A multiple-pronged attack is being used. Banning the extraction of groundwater and natural gas from the earth beneath Venice has slowed the sinking. Steps are being taken to prevent erosion. The MOSES project—a network of mammoth hydraulically operated gates designed to block high tides—is in the works, but it has yet to get off the drawing board after twenty years.

But, as we have seen, it is not the flooding that negatively affects the wood pilings but rather the increased incidence of low tides, which leave the upper heads of the pilings exposed to air and susceptible to rot. Allan Jerbo, a Swedish professor of "the biology of buildings," has developed a system for injecting large quantities of boron—yes, the same ingredient in the Twenty-Mule-Team Borax in which your mother once washed your clothes—into the piles in order to kill the microbes that cause rot. The boron also creates a sort of impenetrable skin around the piling for future protection. All in all, it's estimated it will take until the year 2050 to "boronize" all the accessible pilings—a tremendous task, but one costing an estimated one-tenth the amount of restoring the buildings and foundations.[14]

Perhaps Venice is earth's laser pointer for driving home the point that if man is going to tamper with Mother Nature, he'd better err generously on the side of safety. It is a place like no other on earth. Man created Venice out of water, wood, mud, and faith, and it is this delicate balance that must now be adjusted in order to save her. Meanwhile, the ancient larch, alder, spruce, and oak that have proudly carried Venice on their shoulders for fifteen centuries continue to quietly do their job.[15]

WOOD PIPE TAKES A BOW

Halfway through my visit with wood anatomist Regis Miller of the Forest Products Laboratory in Madison, he opens a glass-front curio cabinet and begins pulling out, well, curios. He pulls out a well-used two-fingered bowling ball made from a solid hunk of lignum vitae. He withdraws a tube of Peelu, an ancient type of tooth whitener derived from the *Salvadora persica* tree, which you can buy today at Whole

Two sections of 120-year-old wood pipe unearthed in Holly, Michigan, in 2007. In Detroit, 50 miles away, over 22 miles of wood pipe have been unearthed.

Foods stores. He pulls out other things that have been in a state of curio so long that no one was quite sure why they were curios.

But surely the oddest things he pulls from the cabinet are three sections of wood pipe. Not the kind your grandfather smoked on the front porch, but water pipe. The three pieces are 6 to 8 inches in diameter, with a 2-inch hole bored down the center. One is a piece of elm with an attached note saying it had come from England and been in service for two hundred years. Another piece of oak, with bark still intact, is simply labeled "Cincinnati, Ohio," while the third has "Detroit—118,000 feet" (another way of saying 22.34 miles) jotted on it with a fine felt-tipped pen.

When one pictures the wood infrastructure of a city, one imagines massive timber bridges, pilings, and towering buildings—not water pipe. But Jon Schladweiler, historian of the Arizona Water and Pollution Control Association, will set you straight. "Log pipe was used all over. In Europe, England, Norway, Sweden, Denmark, Canada, America—any place there were trees."

Wood wasn't used merely for water pipe but for sewer pipe, irrigation pipe, industrial spillways, flumes, conduit for electrical, telephone, and telegraph wire, and even for low-pressure natural gas pipe. Nor was it used only on a small-scale basis or in the foggy distant past. There's a chance your grandfather brushed his teeth using water carried by wood pipe; there's even a chance he's doing that today. Schladweiler, who travels extensively around the country exhibiting his Sewer History Display, says, "I've had people from communities all around the country—East Coast, West Coast, Midwest—tell me they have wood pipe still in service. It's primarily in outlying smaller areas, but it's in the ground and it's functioning well." As late as the 1960s, the not-so-small area of Sedona, Arizona, had sections of log pipe in its water system.

Wood pipe took two basic forms. Small pipe, with an interior bore of 2 to 6 inches, was normally made of long, straight, solid logs with the center bored or burned out. Larger pipe and pipelines, with diameters ranging from 6 inches to 16 feet, was made using barrel stave construction.

Just how durable was log pipe? It was so long lived that the aforementioned wood pipe found in Sedona had previously been installed in Jerome, Arizona, as part of a mining operation; it was dug up and reinstalled in the mid-1930s. Another source discusses 400 miles of elm logs installed in London in 1613 that, when unearthed in 1930, was found to be perfectly sound.[16, 17] In 2004, construction workers in lower Manhattan unearthed several sections of the city's early wood water mains. The sections, made of yellow pine logs, were used to distribute water from a reservoir just north of Chambers Street, an area six blocks from the former World Trade Center towers. The sections were slightly tapered so the narrower end could fit into the wider end. Metal bands were used to secure sections to one another. Houses containing five or fewer fireplaces paid $5 a year to tap into the water service; $1.25 per year per fireplace was added for larger homes.[18]

Log pipe was used at the Neshaminy Water Treatment Plant in Pennsylvania until 1909; while in service it conveyed 13 million gallons of water per day. Log pipe was used for water mains in Philadelphia; Boston; Hartford, Connecticut; Lansing, Michigan; Australia; and London. When it dawned on early firefighters to use these water mains when fighting fires, they began carrying special augers. They would dig down, find the log pipe, and then bore a hole into it. Water would fill the excavated hole, creating a "wet well" from which firefighters could pump water or dip buckets for bucket brigades. When the fire had been doused, firefighters would drive a redwood plug into the hole and mark its location above ground, so the same plug could be used for fighting future fires. Thus arose the term "fire plug." Legend (perhaps urban legend) has it that the small hook on the end of firefighting hatchets was designed for pulling plugs from wood water mains.

The earliest wood water systems were rather crude affairs. During colonial days, 3-foot or 4-foot lengths of pipe were augered out by hand, then joined together using bands of leather. As the process became more mechanized, logs were hollowed by clamping them to sliding tables, then pushed into a long auger driven by a water wheel. Eventually an entire manufacturing industry grew up around wood pipe. The California Redwood Pipe Company's slogan was "Wood pipe is good pipe," and the company promoted it as "a pipe the water companies use and recommend."[19] In addition to the tapered joining system employed in New York, some water mains used a type of ball-and-socket joint, others a wood hub and sleeve, and still others wide metal bands or other devices. As water flowed through the pipe the wood expanded, creating a watertight joint. Often the logs were spiral wound with steel strapping so they could withstand higher internal pressure. The best pipe was also coated with asphalt to help preserve both the wood and the iron banding.

When cast iron pipe came into common use, wood pipe makers, rather than folding up shop, took the opportunity to compare the benefits of their well-made log pipe with the competition. Wood log pipe was cheaper to buy and install; easier to tap for service connections; less liable to freeze, since it was an excellent insulator; and less likely to burst, because of its elasticity. It was claimed to be more durable and better at maintaining the purity of water that flowed through it. One particular type of pipe was positively reviewed by an early building magazine that said, "no section has ever been known to burst so badly that a fire pressure of 80 pounds to the square inch could not be kept on the line of the pipe."[20]

Wood pipe was not without its flaws. After the Civil War, many western communities began importing log pipe for bringing water down from the mountains into town. "I guess it's no different than today," explains Schladweiler. "Everybody had to have their thrills, and cowboys would come along and shoot holes in the pipe just to see a fountain of water shoot 10 feet up into the air. Sometimes lots of fountains. I guess they thought that was great fun. But the great thing was whoever was in charge of fixing the pipeline could just whittle a little plug out of wood and bang it into the hole and the plug would swell up and the pipe would be fixed. You can't do that with metal pipe."

Perhaps even more magnificent than log pipes were the Paul Bunyanesque wood-stave pipelines constructed for transporting large volumes of water for industrial applications. As sawmills became smaller and more portable, logging operations could be set up anywhere—but many still required water to power them. When a river couldn't bring power to a remote sawmill, the river was brought to it. Large "half-moon" wooden flumes were constructed to bring water down from mountains and lakes to drive the saws. Once it had finished spinning the water wheel, the water was often used to transport logs and lumber farther downhill in more flumes.

Some of the largest wood pipelines were used for penstocks for diverting water to turbines generating hydroelectric power. In 1918, the Montana Power Company built a pipeline 14 feet in diameter. In 1934, a wood stave penstock 16 feet in diameter was built by the California-Oregon Power Company. These behemoths were assembled using rows of staves with staggered seams, cinched tightly together every few feet with metal hoops or bands.

Octagonal wood culverts became popular when metal was diverted to the war effort in the 1940s. Short interlocking sections of wood were joined end to end and held together with wood dowels. Over 100,000 feet of octagonal wood pipe—some up to 36 inches in diameter—were used for culverts and storm sewers in 1942 alone. It was not only economical to build but economical to ship: 1,500 feet of wood pipe weighed 35 tons and required one flat car to ship. An equal amount of reinforced concrete pipe weighed 225 tons and required ten flat cars. Wood pipe was downright patriotic.[21]

Naturally rot-resistant woods like redwood and Douglas fir were often employed, but tamarack, hemlock, oak, elm, spruce, and other available woods were also widely used. Schladweiler says he's even seen wood pipe made from palm tree trunks. The durability and longevity of a pipe didn't depend as much on the type of wood it was made from, as on the environment in which it was used. Pressurized wood pipe that continuously carried water could, and would, last for centuries. Because the pipe was constantly wet and constantly surrounded by damp soil, oxygen was denied and the wood became immune to deterioration.

Wood sewer pipe and pipe that carried water or other fluids on a cyclical basis were not as long lived. The alternating cycle of moist-dry-moist-dry set up the ideal conditions for rot. Deterioration was further accelerated by the corrosive nature of sewage or industrial waste. So as cities (and engineers) sought more certainty, wood began to be used less and less, while concrete, cast iron, copper, and steel were used more and more. But wood, as in so many other cases, served as the faithful prototype for these later replacements.[22]

BUILDING A STAIRCASE TO HEAVEN

If you walk through the plaza of old Santa Fe, New Mexico—past the Sleeping Dog Tavern, past the scores of jewelry and pottery stores, past the Cowboy Legends store where you can fondle a pair of Gene Autry's boots—you'll find the Loretto Chapel. It's a chapel that, by chapel standards, most would call modest. Less than 26 feet wide, it has a small classic rose window, the Stations of the Cross, and twenty-one oak pews.

Yet, this seemingly unremarkable chapel, on a good summer day, will attract fifteen hundred visitors from around the world, each paying $2.50 to see the miracle tucked along one side of the church. They don't come to pay homage to a sweet potato bearing the likeness of Mother Teresa or the bones of a saint. They come to see a staircase—more specifically, the "Miraculous Staircase": a wooden double-helix stairway that makes two full 360-degree turns with no visible means of support, built in 1878 by an unknown carpenter.

The belief that the staircase is miraculous is based on three unknowns: Who built it? How was it constructed? What wood was it built from?

The "who" part of the mystery is based on a long-standing legend. The Sisters of Loretto—a teaching order of nuns that moved from Kentucky to Santa Fe in 1852, prompted by Bishop Baptiste Lamy's dilemma and plea that he had "6,000 Catholics, but only 300 Americans"—pooled their finances, some $30,000, to have a chapel built. The sisters called upon St. Joseph, the patron saint of carpenters and builders, to guide them through the five-year construction process.[23]

When it was realized that, because of space constraints, the only access to the choir loft would be via a ladder—a 20-foot ladder, quite difficult to navigate by someone wearing a full-length habit—the sisters began looking for a solution. But architect after architect and builder after builder told them there was simply no way to construct a staircase without taking up a disproportionate amount of space within the confines of the small chapel. The sisters began a novena to St. Joseph in quest of a solution. On the ninth and final day of prayer, an elderly gray-haired carpenter, leading a donkey and toting a tool chest, knocked on the convent door looking for work. Their prayers had been answered. He was hired, and over the course of the next six to eight months, using only a hammer, saw, chisel, and T-square, he constructed the miraculous spiral staircase. Then, just as quickly as he had come, he vanished, never asking for payment or even leaving a lumber bill.

Who was this masked man? Bill Brokaw, who serves as the all-in-one curator, ticket taker, and greeter of the Loretto Chapel, shakes his head and explains, "No one really knows. Right now there are thirty-two claimants who say one of their ancestors did it. The claims are usually based on someone running across a piece of paper saying

The Miraculous Staircase in Santa Fe, New Mexico, built by an unknown carpenter using only simple hand tools in 1878, climbs 20 feet from floor to ceiling with no visible means of support.

their great-great-grandfather built a stairway somewhere in Santa Fe 125 years ago—but there are lots of old staircases in Santa Fe." Some attribute the construction to "Sawdust Charlie," a carpenter, gambler, and friend of Billy the Kid; others to Jose Rodriguez, who was making good on a promise that he would work for the sisters if they reformed one of his sons to Catholicism; and still others to Francois-Jean Rochas, a master carpenter turned recluse.

Carpenters of French, Austrian, German, and Mexican descent have all, at various times, been credited with building the staircase. Of course, there are many who believe that the true claimant is one of Jewish descent: St. Joseph himself. In his book *Loretto Chapel: The Miraculous Staircase,* Brokaw writes, "Saint Joseph, the Patron Saint of Builders and Carpenters, as the builder of this special staircase has been whispered, believed and proclaimed by the believers whose hearts beat faster at the sight of this miraculous staircase."[24] This view by Brokaw, a retired Methodist minister, leans more toward the ecclesiastical than the empirical, but clearly Brokaw isn't the only one who feels it was *the* St. Joseph who arrived in Santa Fe swinging a hammer that day. Many view it as a place of miracles. In the short time I was there, dozens of visitors reached across the velvet rope and, with head bowed, caressed the closest tread and said a prayer. In the past, many visitors were much more vigorous in paying homage; in their fervor to possess a keepsake from the Miraculous Staircase, many would routinely chip off slivers of the old horsehair plaster that coated the underside of the staircase. To put a stop to it, when the bottom was replastered, it was faux painted to look like wood boards in an attempt to deter iconic shoplifting.

The "how it was built" part of the mystery can be studied a bit more pragmatically. Here are the hard cold facts. There are no center or outer perimeter support posts of any kind. The staircase possesses an uncanny resemblance to a classic DNA molecule standing on end. The stairway is built of ninety-three pieces of wood, climbs 20 feet from floor to balcony, and consists of thirty-three treads: coincidentally or not, one for each year Christ lived. The stringers—the inner and outer supports on which the treads rest—are composed of eight and ten pieces of wood, respectively. No single piece of wood exceeds 5 feet in length, and the multipiece stringers are secured to one another using simple half-lap joints and wooden pegs. No nails, screws, laminations, or glue were used. The staircase was originally built without handrails and spindles, and it was used as such for seven years until they were later added.

Truth be told, it does appear to defy the rules of physics, carpentry, and logic. It is a thing of solidity and grace, gentle in its ascent and ribbonlike in its construction. It carries the glorious patina of a wooden artifact that has been lovingly rubbed, trod upon, and used. The view from the top shows such perfect symmetry that viewers might think they were gazing at the top of a well-poised conch shell.

"There have been eleven tries to replicate this stairway, and none have succeeded. People have tried building it laying down then standing it up and every other which way, and no one's even gotten close," Brokaw explains. "There are eight carpenters out of Longview, Texas, who came here and studied the staircase for two days, and they're trying to replicate it now. I'll be curious how it turns out." Brokaw says this with the gleam in his eye of someone who already knows the outcome.

Three earthly based (and one heavenly based) theories try to explain how the staircase "works." The first proposes that the inside stringer, being of such a tight diameter, actually serves as a load-bearing column. But engineering studies have discounted this on the basis that the diameter is not actually tight enough. The second theory views the staircase as a gigantic spring that ever so slightly compresses when in use, then returns to its original shape and height again. But after 125 years of use, with tens of thousands of trips taken up and down them, with entire choirs sometimes assembled along its spiraling perimeter, the stairs haven't compressed a whit. The third theory revolves around the joinery. Some feel the staircase—with treads, risers, and stringers all solidly joined to one another with square pegs and exquisite craftsmanship—create a single, solid entity. The theory is that the pegs are positioned in such a manner that they either do their job or shear off—and there's no in between; thus far, they are doing their job. But all attempts to replicate the staircase using this method, even with modern tools, have ended in failure. Brokaw posits, "Could it be a combination of all of the [above] therories, with a large portion of prayer and the will of God? Despite these theories, no absolute consensus among engineers has ever been reached. Many say that it just cannot be explained."

There are other construction-based mysteries. How did a carpenter, working alone and with minimal tools, curve the stringers and do so with such accuracy? Some nuns reported seeing boards being soaked in a large barrel of water—but recent analysis of the wood fibers points to steam-bending versus wet-bending, and steam-bending would have required a much more complicated apparatus than a barrel. How did a single person, without benefit of blueprint or trial and error, build a perfectly symmetrical staircase? And where did the wood come from?—which leads us to the third great unknown.

In 1996, Forrest Easley, a native New Mexican with a BS degree in forestry and wood technology and years of experience in the fields of timber management and research, conducted an analysis of the wood used to build the Loretto Chapel staircase. His written analysis is prefaced by this statement: "During my formal training as a scientist, it was impressed upon me very early that one absolutely must eliminate all personal feelings and bias from a study in order that it be truly a scientific study. So in that light, this work was carried out."[25]

And what did Easley find? On the basis of a small sample taken from the upper inner stringer, he determined that the wood was a close relative to Sitka spruce and Engelmann spruce. But there were oddities. For starters, neither wood was available in the immediate area, at least not in the sizes used in the staircase. What's more, the tangential wood grain—what we might call the face of the board—was wavy, not dimpled or plain as in other spruce. And the longitudinal tracheids—the things we might think of as the long, skinny cells that run parallel to the tree's length—were "more or less square," unlike any other known spruce. Which led Easley to conclude, "Therefore, I, the investigator, since this specimen is of a wood heretofore unknown, hereby assign to the thus-far unnamed species within the genus Pinacae Picea the following names: Suggested scientific name:

Pinacae Picea josephi Easley ["josephi" for St. Joseph, "Easley" for the researcher]. Suggested common name: Loretto Spruce." Allegedly, no other wood of this exact species has been found since.

When I conferred with Dr. Regis Miller—who for most of his forty years at the Forest Products Laboratory in Madison headed up the Wood Anatomy Research department—I could almost hear his eyes roll over the phone. "If you're going to name a new species, you have to go though this huge process. You can't name a new species based only on the wood—you have to do it based on the whole plant. The woods from most spruce [and there are ten species in North America alone] look the same. You have to look at the needles, cones, and other herbaceous material to really differentiate a new species." Miller went on to explain that the description needs to be written up just a certain way in Latin, that the findings need to be published correctly, and that there has to be acceptance by the broader botanical community. Miller explained, "He used a lot of 'therefores' and 'herebys' to make it all sound official, but he didn't even spell 'Pinaceae' right."

Miller sent me an example of paperwork recently submitted to the American Society for Plant Taxonomists for a new species using the correct procedures and nomenclature. It's ten pages long, complete with illustrations of staminate flowers and stem cross-sections, charts of diagnostic characteristics, and micrographs of pollen. It *does* have lots of Latin words, and it's absolutely littered with sentences like "Serjania can be distinguished from most other genera of Paullinieae (sensu Acevedo-Rodriguez 1993a; i.e. *Cardisospermum, Houssayanthus, Paullinia,* and *Urvilleas*) by its schizocarpic fruits with winged medicarps and distal cocci."[26]

What does Miller think the staircase is made of? "For at least a hundred years, people have been dragging plants and seeds all over the place when they move, so it's hard to even pin down what kind of wood you have based on where you find it." Foxtail pine, a rare and long-lived relative of the bristlecone pine, is the first wood that springs to Miller's mind; pinyon pine is another. Miller was turned down ten years ago when he requested a small sample of the staircase to examine; he'd still like to examine a sliver or two.

But even if you disregard the "heavenly wood" part of the mystique, the staircase remains remarkable on the other two counts—who made it and how it was made. And, as Meatloaf reminds us, "Two out of three ain't bad."

Other miracles made of wood are found in churches around the world. In Wieskirche, Germany, in a church containing some of the finest rococo architecture in the world, you can visit the statue of the "scourged saviour." The statue was originally made early in the eighteenth century by piecing together parts of several other statues, binding the joints together with strips of linen, and repainting the whole thing. The end result—which at best resembles a papier-mâché mannequin with leprosy—was carried in the Steingaden Good Friday processions in the 1730s. But eventually the congregation became so upset by the statue's pathetic appearance that it was hidden away in a monastery attic. When the statue was being moved to a farmhouse in 1738, the owners saw tears in the statue's eyes—and the legend of the scourged saviour was off and running. People flocked to see the statue. A smaller, then larger, then fantabulous church was built. Today thousands of people still make the pilgrimage every year.

The mahogany statue of Our Lady of Fatima also allegedly cries human tears. She's visited more than a hundred countries and has been credited with curing blindness, cancer, and other terminal illnesses like atheism. Carl Malburg, who has been caretaker of the 40-inch-tall statue for thirteen years, maintains that scientific analysis in 1972 showed they were real tears. Malburg says that she's cried thirty times, but adds, "the last time we have photographic evidence was from 1995 in Indiana."[27] Skeptics cough, scoff, and wail, "Haven't these people ever heard of sap? Or condensation?" But you'll never convince the 150,000 people who recently visited the statue in India of that.

The Miraculous Staircase may well be the world's most famous set of steps. It's been featured on "Unsolved Mysteries," in "Ripley's Believe It or Not" column (twice), and on *Good Morning America*. There have been documentaries galore, and it served as inspiration for the movie *The Staircase*, starring Barbara Hershey. It's been designated as one of New Mexico's Seven Great Wonders.

But why the appeal and fascination? One can only hazard a guess. Perhaps it's because sculpture can stir one's emotions, religious artifacts can touch one spiritually, well-engineered items can appeal to one's intellect—and the Loretto staircase seems to do all three. If you're hitting on all three cylinders, you can't help but feel moved by the structure. Stone-cold realists can rightfully maintain that the staircase was built from everyday spruce by a masterful carpenter with no business sense who happened to be in "the zone." Snake-handling Christians can elect to believe that the staircase was built by St. Joseph from a wood milled in heaven, with an integral part of its structure composed of divine intervention. Or you can be anywhere in between. That's the beauty of the staircase.

The staircase has had its ups and downs. Several years ago, the floor beneath it rotted out, and the staircase sank 6 inches. Some surmised that if St. Joseph was attentive enough to build the staircase in the first place, he would surely show up to patch the floor. But apparently saints prefer building over remodeling; after a long wait, a concrete pad was poured to remedy the problem. A small metal plate has been installed to reinforce one of the splices in the inner stringer. It's been years since the general public has been allowed to use the staircase; however, if you reserve the chapel (now under private ownership) for a wedding, the bride and groom are allowed to have their picture taken on the lower part of the staircase. But for the most part, people come and wonder.

So where does it all come down? As I leave I ask Brokaw straight out: "Do you really think we're dealing with a miracle here?" And he responds matter of factly, referring to the nuns' answered prayers of 150 years ago: "Any time a prayer is answered it's a miracle. And that's a pretty good deal, don't you think?"

ACADEMY AWARD NOMINEES FOR OUTSTANDING PERFORMANCE BY A WOODEN STRUCTURE

With typical authorial zeal I began researching and formulating the case for wood being the most fabulous building material of all time. After days of research I began realizing how little had been written on this wondrous fact. Why such scant information about the wood equivalent of the Great Pyramids? The Parthenon? St. Peter's Basilica? And then I realized the problem. The problem was that my fact wasn't a fact. Wood is good, plentiful, and easy to build with, and, when kept very dry or very wet, it will last thousands of years. But in the end, wood is wood—and the elements know it. Metal may rust, but it doesn't burn. Stone may erode, but it doesn't rot. Brick may slowly deteriorate, but there's not an insect on the planet that likes it for supper. Not one of the Seven Wonders of the World was built of wood.

In *The Wood Book*, Jan Adkins states, "Time, the lord of decay, is kind to well-built wood structures. Some of our oldest buildings are wooden. Japan's Golden Hall was built in AD 679, survived a major fire in 1949, and is still open to the public. Scandinavian log halls and dwellings from the year 1000 are still robust and sound."[28] But alas, this isn't a very long sentence, punctuated with comma after comma, is it? No, in terms of size, scale, and endurance, other materials are often better suited for structures. So, with tempered enthusiasm, I head through the hallowed halls of history—halls built of stone—to find the Oscar nominees for outstanding wood structures.

Nominees for Oldest Wooden Building. Dozens of Japanese temples have survived intact over the past thousand years. Unsurprisingly, most of the structures are in Buddhist monasteries, where they've been regularly maintained and been kept out of the tourist fray that can account for so much wear and tear. Indeed, most of the oldest wooden buildings in the world are religious structures. A wooden stave church near Oslo, Norway, dates back to the thirteenth century, and some wooden Slovakian churches were built as far back as the sixteenth century.

* * *

Two buildings within the Horyuji temple complex near Osaka, Japan, are considered by most to be the world's oldest wood buildings. The Kondo Hall was built in 607. Records indicate that it was struck by lightning and burned to the ground in 670, rebuilt in 700, then repaired and reassembled at least three times—in the early 1100s, in 1374, and in 1603. It's estimated that only about 20 percent of the original temple remains. The adjacent five-story pagoda was built in 700.

The Shrine of the Sun Goddess, a Shinto shrine in Ise, Japan, was built in the third century and looks exactly as it did eighteen hundred years ago. However, this amazing longevity isn't a testament to the enduring quality of the Japanese white cypress from which it's crafted, but rather a testament to the enduring tradition of tearing down the building every twenty years and rebuilding an exact replica on an adjacent site. This, according to tradition, not only purifies the site but also supports the Shinto belief that nature doesn't make monuments but lives, dies, and is continuously renewed and reborn.[29]

Beams from the Jokhang Temple in Lhasa, Tibet, indicate that the temple was built during the seventh century. The 555 growth rings found on one of the massive beams indicate that the tree it came from germinated in the first century.[30]

The Sakyamuni Pagoda in northern China would most likely win the combined award for the oldest and tallest building. Built in 1056, it stands (albeit slightly lopsidedly) 216 feet and nine stories tall.[31]

There is evidence of much, much older wooden structures. In an area north of Tokyo, archaeologists found a site where ten holes were dug in a pattern, 500,000 years ago, presumably to hold posts for supporting walls and a roof.[32] Ruins at Terra Amata near Nice, France, dating as far back as 400,000 years ago point to similar wood structures. But alas, no actual wood was found, only clues that it was once there.

Nominees for BIGGEST Wooden Building. "Big" can be defined in many ways. It can refer to how much lumber a building contains, how much surface area it covers, how big the roof is; it can even be a sentimental favorite. The Great Buddha Hall in Nara, Japan (which shelters an absolutely humungous 250-ton statue of Buddha—one that according to legend took two and a half million people and nearly all of Japan's bronze production for several years to build), is considered by many to be the world's largest wooden building. Its original size was 180 feet in width, 288 feet in length, and 154 feet in height. The temple was rebuilt twice after fires, and the one now standing is 30 percent smaller than the original.

A wooden blimp hangar built for the U.S. Navy during World War II, and still standing, measures 300 feet wide, 190 feet tall, and over 1,000 feet long. The clear span is 5.6 acres, and if you spread the roof out flat, it would cover 11 acres. It took 2 million board feet of fire-retardant Southern yellow pine timbers measuring 3 × 8 inches to 6 × 14 inches to build it. Coffee rationing during the war apparently had little effect on the workers' energy levels; the main structure was erected in twenty-seven working days.[33] Some withhold the "World's Largest Wood Structure" designation to this hangar because it's more of a gargantuan garage than a building. Its equally large twin, which burned to the ground in 1995, would surely be a candidate for "World's Largest Fire"; flames could be seen from 50 miles away.[34]

Some feel that the "World's Biggest" award should be handed to the Odate Jukai Dome in Odate, Japan, which measures 584 feet by 515 feet and required 25,000 Akita cedar trees to build.

Others claim that the Woolloomooloo wharf and warehouse near Sydney, Australia, built in 1912, is the largest wooden structure. It measures 1,312 feet long and 206 feet wide and stands on 3,600 wooden piles. The math indicates while it may not be the largest, it is perhaps the longest.

Nominees for Tallest Wooden Structure. The tallest wood structure in the world is an aerial tower built in Germany in 1934 that stands 364 feet high and is built of larch, with brass screws used for the connections. Appearing a bit like an anorexic Eiffel Tower, the structure, according to scientists, still has another twenty years of life left in it. That's pretty tall, but you'd need to stack four and a half of these towers atop one another to equal the 1,667-foot height of the Taipei 101 building in Taiwan.

This wood World War II blimp hangar stood 190 feet tall, 300 feet wide, and 1,000 feet long. It was built from 2 million board feet of Southern yellow pine in twenty-seven working days.

While wood hasn't created the tallest structures on the planet, it has created a sense of place in our hearts. Even when a ninety-story concrete and steel building is erected in the center of New York City, the tradition is to top it off with a tree when it nears completion. Some call it "wetting the bush" as a reference to the toast that was traditionally raised at the same time the tree was.

Nominees for Longest Wooden Bridge. There are many claimants for the longest bridge award. The wooden Mon bridge in Thailand, 2,950 feet long, lays claim to the world's longest pedestrian suspension bridge. The 1,282-foot Hartland bridge in New Brunswick, Canada, lays claim to longest covered bridge. Other bridges are designated as the longest free-span bridge, the longest pontoon bridge, and the longest toothpick bridge.

The Mile-Long Bridge in Hampton, New Hampshire, which opened in 1902, is considered by some to be the world's longest, even though it should have been more accurately called the 7/8th-Mile-Long Bridge. But you can't blame a community for rounding 4,700 feet up to an even mile for posterity—and to better fit on a sign. It stood on 4,000 (or 5,270 or 3,380, depending on what you read) oak piles, each 28 feet long, cut from surrounding forests. It required 1.8 million feet of pine and 25 tons of iron bolts to build.[35] The decking was also made of wood. It was sturdy, carrying train traffic on one edge, pedestrians along the other, and vehicles in between. The wooden bridge had its share of problems. Pilings sank, storms washed away a 600-foot section, and 125 feet of it burned, but repairs continued to be made. The cracks between the wooden decking planks provided a secure resting place for discarded cigarettes to nestle and burn. In one ten-week period, firefighters were called fourteen times to extinguish bridge fires. It was replaced in 1949 by a new-fangled concrete and metal version.

Maybe the coolest wooden bridge was Trajan's Bridge—the first bridge ever to be built over the Danube. It consisted of twenty wooden arches, each spanning 170 feet, set on stone pillars, which were built on oak or alder piers driven into the river bottom. The bridge was 3,720 feet long, 50 feet wide, and 60 feet above the Danube at its highest point. This is made all the more amazing by the fact it was built in less than three years, between 103 and 105 AD. Some surmise that this was the first time the concept of a truss with triangular bracing was used. We have a rough sense of what the structure looked like because it is depicted in relief sculptures that survive today.

Charlemagne had a bad experience with wooden bridges. He spent ten years building one across the Rhine River at Mayence, only to have it destroyed by fire in three hours.

Nominees for Structure Containing the Most Wood. In 1902, the Southern Pacific Railway built a 12-mile-long trestle, called the Lucin Cutoff, across the Great Salt Lake in Utah. The project resulted in "Forty-three miles in distance [being] lopped off, heartbreaking grades avoided, curves eliminated, hours of time in transit saved, and untold worry and vexation prevented."[36] All said and done, it took 32 million board feet of lumber to stop the vexation.

The project commissioned the construction of twenty-five massive pile drivers, each with 3,200-pound hammers, to drive the twenty-five thousand Douglas fir piles into the sometimes mushy, sometimes rock-solid bottom of the lake. Some piles exceeded 125 feet in length. All in all, nearly 2 square miles of Oregon forest were cut for the pilings alone. In explaining the sheer magnitude of the project, one source explained, "Just the portion of the trestle above the waterline contained enough wood to lay a boardwalk four feet wide from Boston to Buffalo."[37]

The trestle was replaced by a solid fill causeway in the 1950s. In the 1990s, the structure that took three thousand men two years to build was dismantled over the course of seven years. Cannon Structures, Inc., and its Trestlewood Division, which obtained salvage rights to the structure, estimates the trestle contained 10 million board feet of Douglas fir timbers, 20 million board feet of Douglas fir pilings, and 2 million board feet of 3-inch-thick redwood decking, much of which now breathes new life as flooring and timber frame structural members.[38] Trestlewood's motto? "They don't make wood like they used to."

Nominees for Other Really Big Wooden Things. Wood is, and always has been, readily available, easy to work, and inexpensive, making it the material of choice for one-offs, make-dos, and absolutely gigantic things ranging from rolling pins to xylophones.

The city of Troy, Pennsylvania, once hosted the world's largest water wheel, with a diameter of 60 feet and width of 22 feet; it is said to have inspired George Ferris to build a certain large amusement ride. Kalama, Washington, is home to the world's tallest totem pole; the 140-foot icon was carved from a seven-hundred-year-old red cedar.

There are, of course, many small wooden things that have been made on a grand wooden scale. The largest wooden golf tee stands 10 feet 9 1/4 inches tall. The largest hockey stick measures 110 feet in length and weighs 5 tons. The largest yo-yo was made by the woodworking class of Shakamak High School; it was "yo-ed" using a crane and yo-yo-ed twelve times. The world's largest pencil, 65 feet tall, stands outside the Faber-Castell pencil factory in Malaysia. The world's largest wood baseball bat is actually made of 34 tons of steel, but it *looks* like wood. It stands 120 feet tall and casually leans against the five-story Louisville Slugger headquarters. The world's largest chair is 20 feet 3 inches tall, weighs 7,900 pounds, and is made of solid ash. It's a good place to sit while pondering the wonders of wood.

ROLLER COASTERS: MÖBIUS STRIPS OF SCREAMING WOOD

Want some heart-pounding roller coaster facts? The fastest—Kingda Ka at Six Flags Adventure—hits 128 miles per hour. The longest—the Daidarasaurus in Expoland—has a length of just under 1 1/2 miles. And the steepest—Speed at Oakwood Leisure Park—has a descent that's a physics-defying 97 degrees. But beyond these hard, cold statistics you'll find discord in the roller coaster world, and nowhere is the debate more heated than when the discussion turns to the matter of which provides the most thrilling ride: wood coasters or steel?

According to Ken Felber, who is overseeing the construction of one of the Midwest's largest wood roller coasters, the Renegade, at Valley Fair Amusement Park, it depends on what you like. "The ride on a steel coaster is like a sports car on a smooth asphalt track. A wood coaster is more like a truck on a logging road."[39]

Steel coaster enthusiasts thumb their noses at old-fashioned wood. They prefer the smoother, faster, corkscrewing, somersaulting, loopier ride that steel tracks and urethane wheels deliver. But hardcore wood coaster traditionalists consider steel coasters too sanitized. Woodies, they maintain, shake you around, provide more air time, and are more fun because their movements are more fickle and less predictable. Felber thinks there's more to it than that. "Lots of people like the aesthetics of a wood coaster. They're just so massive. And part of it is nostalgia. They even sound different. The feel of the ride is completely different."

Whichever coaster car you reside in, you owe your twenty-first-century adrenaline rush to a group of seventeenth-century Russian thrill seekers—Catherine the Great among them. Russian ice slides were built of wood, looking like mammoth skateboard ramps covered with ice. One would ascend stairs, hop on a sled (sometimes also made of ice), and zing down the 50-degree slope. Catherine the Great was so enamored with the ride that she had slides built on her own grounds (which, according to legend, is not the only way the sexually voracious empress got her ups and downs).[40] But I digress.

It was just a matter of time before someone—either Russian or French; it's debatable—fashioned a track and developed a sled with wheels to ride on it so people could be thrilled and terrified year round. By 1817 there were wheeled roller coasters in France. One could reach speeds of 40 miles per hour—in that day the fastest thing on land, air, or sea the average person could ride on. The first upside-down roller coaster, which included a 13-foot vertical loop, wasn't far behind. And though sandbags, flowers, eggs, and monkeys were used as test pilots before humans were allowed on board, the coaster had an uncanny tendency to snap riders' necks and was quickly banned until safer versions came along in the 1890s.

Roller coastering in America took off in 1829, but at a considerably slower pace.[41] The first coaster was a coal train—the Mauch Chunk Railway—in which people paid a then-hefty 50 cents apiece to zip down the track at a whopping 6 miles per hour in converted railroad cars. Each downhill thrill ride was also enjoyed by a team of mules, which would disembark at the bottom and pull the cars back up for the next trip. As the ride caught on, it was refined. Cars were pulled up a half-mile incline via a steam-driven cable system, then released for a 6-mile gravity-induced coast. The return trip was accomplished in a similar manner. By 1874—with the coal operation diverted to another system of tracks—the Mauch Chunk turned into a full-fledged tourist attraction complete with a restaurant, hotel, and thirty-five thousand visitors a year.

The first bona fide coaster was built at Coney Island in 1884, and it was downhill for the amusement park industry after that, at least for a while. Much of the roller coaster craze was stimulated by trolley companies that—lacking adequate ridership on Saturdays and Sundays—built amusement parks near the ends of their rail lines on the outskirts of town to increase weekend revenues. The Golden Age of Roller Coasters ensued. In 1929, things peaked, with fifteen hundred coasters worldwide, including The Cyclone at Coney Island, which yet today remains an industry icon. Since that year the industry has weathered its hills and valleys. During the Great Depression, ridership sank and coasters fell into disrepair. By World War II, the number of coasters in America had dwindled to around two hundred.

Disney's Matterhorn Bobsled Ride gave the industry a shot in the arm in 1959. Subsequent theme parks like Six Flags Over Texas, and the advent of steel track coasters capable of corkscrews and other daredevil feats, put roller coastering back on an uphill track. Suspended coasters, where riders hang from an overhead track; stand-up coasters; and trackless "trough" coasters continue to inject new life into the industry. Propulsion mechanisms similar to those used to launch jets from aircraft carriers and electromagnetic linear accelerators have pushed the speed limits. While the newfangled metal coasters renewed interest in roller coasters and gave woodies a run for their money, wood held on for the ride. In 1979, the wooden Beast—still the longest wood coaster in the world—was built. In the early 1990s, the Georgia Cyclone, the Mean Streak, the Thunder Run, and the Texas Giant wood coasters were built. Several international companies specialize in wood coasters and continue building them at a steady clip today. Wood is alive and well.

On the day of my visit with Felber at Valley Fair, the wind chill is hovering around 5 below, but in true Minnesota-nice fashion he grits his teeth, bundles up, leads me outside, and walks me through the steps involved in building the $6.5 million Renegade.

It all starts with a solid base. The legs of some coasters rest on individual piers, but the Renegade rests on a half-mile-long slab of reinforced concrete that twists and bulges to accommodate the structure. There are 505 vertical support legs. "We call them bents," explains Felber. "It's a term originally used to describe railroad trestle supports. And it's a good term to use, since these roller coasters are basically just big, long trestles."

Each of the 505 bents has its own computer-generated blueprint that specifies every detail, including the angle cut and length of each member, "down to 1/16th of an inch," says Felber. It includes the top member, some of them banked at an 80-degree pitch. The crews build the bents on gigantic assembly tables, then stand them up one by one, using cranes. Big cranes. Most bents are spaced only 3 to 4 feet apart. The largest are over 100 feet tall and 40 feet wide at the base.

There is no sissy lumber in a wood roller coaster. All half-million board feet of lumber going into the Renegade will be 4 × 6s, 4 × 12s, 2 × 10s, or equally hefty timbers. Treated Southern yellow pine is being used, but other rides at the park have used Douglas fir. Each has its

Building a roller coaster requires being half carpenter, half mountain goat, and—in a windchill of 5 degrees below zero—a tough mountain goat. Each of the two track ledgers is made of six two-by-tens laminated atop one another.

tradeoffs. Douglas fir is stable and straight, but it doesn't accept pressure treatment as well and must be punched with thousands of little incisions in order to better absorb the preservative. Southern yellow pine takes in the solution much more readily, but it tends to twist and warp unless you get it fastened down quickly. And these guys *are* working quickly. They use gigantic nuts and bolts to hold the big members together; 16-d nails to hold everything else.

Once a bent is erected and fastened to the concrete, it's secured to the adjacent bent with a horizontal 4 × 8 ribbon. Then more crossbracing is added. When a stretch of bents has been erected, the crews install the catwalk and handrail that will be used for maintenance. Then the ledgers—one for each track—are installed. Each ledger consists of six 2 × 10s laminated atop one another and secured to the tops of the bents. This massive 2 × 10 sandwich provides a solid surface for securing the final track the cars will ride on. Each of the six layers also helps to progressively smooth out the joints and twists. C-clamps the size of outboard motors are used to coax and plead each 2 × 10 into position. Forcing the 2 × 10s to conform to the track where it rises and curves at the same time is particularly straining. But these guys are pros. They listen to Guns N' Roses and drill and bolt and nail and clamp their way through all nineteen turns and 3,113 feet of track. From a distance, Renegade has a massive grace. It's as close to a mountain of wood as you'll ever see.

There are no "purely wood" coasters. Even the woodiest woody, like Renegade, has a 1/2-inch-thick ribbon of steel attached to the wood ledgers for the wheels of the cars to ride on. These tracks are secured through all six layers of the ledger, using sleigh bolts. And in the traditional manner, the carpenters drill a hole, pound the sleigh bolt halfway in, give it a good sideways whack with a hammer to put a bend in it, and then pound it in the rest of the way. The bend locks the bolt in place and keeps it from turning when the nut goes on.

Renegade (Ride with a Vengeance) is big but not the biggest. The Wild, Wild West roller coaster recently constructed in Germany covers 2 1/2 acres, has a track length of 3,600 feet, and required 936,000 board feet of treated Southern pine. It's put together with 90,200 bolts, 113,000 pounds of nails, and the labor of forty-five full-time carpenters. The American Eagle, another all-wood structure, used a whopping 1.6 million board feet of wood, resting on 2,000 concrete piers and assembled with 60,000 bolts and 30,000 pounds of nails. If you submit a painting bid, make sure to include all 9,000 gallons of paint.

The original coasters were basically modified train cars riding on standard railroad tracks. But as speeds increased and turning radiuses decreased, new types of track and car were needed. The solution was a side-friction coaster, where groups of wheels mounted perpendicularly on the sides of the cars would bump and roll against sturdy railings built at side-wheel height. This kept the cars on the tracks with regard to the side-to-side movements. But as coasters picked up speed, a new system was required to hold cars down on the track while zooming over the peaks. A system of under-friction wheels evolved in which a set of wheels ran along the underside of the ledger to keep the cars anchored.

Today, roller coaster design is a science that takes into account biodynamics, G forces, speed, and human psychology. The track must be banked at just the right angle, based on the coaster's speed, so the G forces created (2 or 3gs in the case of Renegade) are transmitted straight down the rider's centerline. If the speed of the coaster and the bank of the track don't jibe, heads will be whipping back and forth—not necessarily a good thing at 52 miles per hour.

When Renegade is complete, engineers will bolt an accelerometer to one of the seats and flip the switch for the 2-minute ride. With Renegade, they'll particularly keep an eye on the 104-foot twisting first drop—the first of its kind. When the accelerometer returns, it will output a line graph showing G forces and longevity. "They want to see a graph with nice smooth transitions in and out of the turns and dips," explains Felber.

There are two types of wooden roller coasters. Some have wooden tracks and ledgers but are perched on a superstructure made of steel. Others, like Renegade, have wood tracks and wood superstructures. But regardless of the support structure—steel or wood—roller coasters that ride on wood tracks are categorized as wood coasters. Of the 172 wood roller coasters operating today, 125 are located in the United States.[42]

The older ones, particularly those that have been snatched from the jaws of progress and restored, require constant maintenance. But so do the new ones. When Renegade is finally rolling, maintenance personnel with pockets full of wrenches will walk every foot of the track every morning, looking for loose bolts or potential problems. Since wood expands and contracts according to the temperature and humidity, knowing how tight to snug the bolts is something that comes only through experience. The areas of the track receiving the highest stress—those in the tight curves or deep dips—may need replacing every three to five years; other areas will last twenty to thirty years.

On modern steel coasters, a weld is a weld; it's either doing its job or not; there's no give. But wood coasters are built slightly "loose" on purpose; they need to have give in order to withstand the jostling and momentum of the cars. Designers don't want a rigid, brittle frame; they want something that can move and flex. And that's what makes them fun. The survival and revival of wood coasters attest to the notion that the highest technologies do not always offer the greatest thrills.

There is the occasional mishap—a coaster in Germany recently got stuck upside down for an hour and a half at the apex of a loop. There are isolated fatalities, usually caused by rider stupidity. But overall they're safe; government records reveal only one injury per 25 million rides, and one fatality every 450 million.[43] Statistically you have a greater risk of injury playing golf, chewing gum, or folding a lawn chair. "We have lots more accidents on our walkways than on our rides," explains Felber.

In fact, riding roller coasters is good for you. Studies show that riding a coaster fools your body into triggering its fight-or-flight mechanism. It gives your amygdala—the part of the brain that controls emotional responses—a kick in the pants and tricks your organs into releasing epinephrine, adrenaline, and other internal stimulants. There are scientific reasons for all the grins and giggles emanating from the riders skipping down the exit ramps.

When you compare just the numbers, steel beats wood in nearly every category. The tallest steel coaster ascends to 456 feet; the tallest wood coaster to less than half that height. The fastest wood coaster, Son of Beast, at 78 miles per hour, travels 50 miles per hour slower than the steel speed champion. The steepest steel coaster has a 20-degree sharper drop than its steepest wood cousin. Steel coasters have loops, corkscrews, and inversions, while most woodies stick with monstrous descents and curves. Steel coasters tend toward video game–type names: Millenium Force, Hypersonic XLC, Steel Force; wood coasters lean toward old-fashioned Marvel Superhero–type monikers: Beast, White Cyclone, Texas Giant. Yet, wood coaster addicts maintain that the ride of a wood coaster is vastly more thrilling. Often the explanation involves "more air time"; the float you get when the coaster zips over the apex of a hill—something steel coasters don't offer, or at least don't offer with the same pit-of-the-stomach ferocity. Richard Bannister, an Irish coaster fanatic, states, "Wood coasters simply have more shake, rattle, and roll." And a guy who's ridden 886 different roller coasters in twelve countries for a total of 837 miles, enduring 1,783 inversions, should know.[44]

One woodie lover explains his passion: "[Woodies] have what I like best, emphasis on drops, air time, and the free and open feeling caused by lap bar restraints instead of over-the-shoulder restraints. They feel more organic and less rigid and precise than steel. Even when they are poorly maintained and have shuffle, vibrations, and jack hammering, I love them, as I actually enjoy those sensations. I love the wild out-of-control feeling, direction changes, sensation of speed, and violent, abrupt "ejector" air. My favorite seat is always the front, because to me the quintessential coaster experience is about feeling the wind in my face, watery eyes, and seeing nothing in front of me except track disappearing below me."[45]

Adrenaline junkies may get a thrill out of riding wood roller coasters, the devout from the sight of the Miraculous Staircase, tourists from visiting the canals and cathedrals of Venice. But all of us should get a thrill out of the fact that wood exists at all. I sure do.

This last section contains ideas on how we can make sure the thrill of wood lasts for generations to come.

EPILOGUE

Trees—Answers, Gifts, and Ducks in the Wind

Antonio Stradivari's incomparable violins, Howard Hughes's monstrous *Spruce Goose*, Jimmy Carter's handcrafted chairs, Barry Bonds's fearsome bat—this strange and varied assortment of objects share little in common but for one thing: they're made of wood—wood that originated in the forests that still today cover 30 percent of earth's terra firma. These forests can be large or small, natural or managed, located from the redwood forests to the Gulf Stream waters to the Amazon basin. Indeed, a discussion of wood without a discussion of the forests and trees from which it's born would somehow be incomplete.

We *have* taken a brief look at trees from the anatomical, historical, chemical, and structural points of view—but two angles have been skirted: the political and the environmental. These realms, which rile up matters of deforestation, global warming, tree smuggling, and much more, have been largely and purposefully sidestepped. Why?

For starters, this book attempts to present the story of wood—at least in part—without hidden agendas or value judgments. Stories weren't omitted, censored, or slanted because the woods, or the forests from which they came, weren't necessarily used with the best interests of the plant, planet, or mankind at heart. These stories are ecologically neutral.

Second, facts about the relationship between trees, humans, and the environment are far from being cut and dried. One source puts the pace of deforestation at 21 acres per minute, while another puts it at 150.[1] One source puts the amount of CO_2—a greenhouse gas that contributes to global warming—that an acre of trees will annually sequester at 2.5 tons,[2] while another puts that figure at four times that amount. Objective data are hard to come by. The numbers depend not only on your forest and math but also on your politics.

Finally, living trees are a particularly emotionally charged subject. While a thousand board feet of Douglas fir sitting in a pile at ACME Lumber is largely noncontroversial, that same thousand feet standing in the Cathedral Grove can be a flash point. Witness the spotted owl controversy in the 1980s and 1990s, when environmentalists' bumper stickers blared "Save an owl—ban logging," while the timber industry's retort was "Save a logger—eat an owl." Witness the story of Julia Butterfly Hill, who sat in a six-hundred-year-old redwood for two years to protect it from logging, only to have it surreptitiously cut through and through a few weeks after she climbed down, having reached a successful compromise in her mission.

Well, the next couple of pages have to do with the emotional, environmental, and political. Some opinions are expressed here. You've been warned. Just slam the book shut right now and pass it on to your brother-in-law if you don't want to hear about it. Really, that's okay.

During my search for a balanced view of the situation, I talked with Patrick Moore, a founding member of Greenpeace, who served nine years as president of Greenpeace Canada and seven as a director of Greenpeace International. You're thinking, "Oh, yeah, *there's* a real unbiased point of view!" But you need to read the entire biography. Because after leaving Greenpeace, Moore began shifting his energies more toward sustainability, consensus building, and the big picture.

When asked about the relationship between people, trees, and the environment, Moore explains: "[some] assert we should cut fewer trees and use less wood to save the forest. I believe that is incorrect. In fact it is antienvironmental, because when you use less wood you automatically use more steel, concrete, and plastic. And the manufacture of these nonrenewable resources are part of the cause of CO_2 emissions in the first place." Moore continues, "So my belief is the correct policy could best be summed up as 'grow more trees and use more wood.'"

Wood and trees play a much more critical role in the world than serving as the raw materials for the pianos, pencils, and chain-sawn bears examined here. The estimated 8 billion to 10 billion acres of forest on this planet have a much more serious, businesslike side.[3] Trees do things, like these:[4]

Trees sequester carbon dioxide and in doing so help reduce the greenhouse gases that contribute to global warming. Forests absorb about one-fifth of the world's carbon dioxide output.[5] To put it in another light, a healthy forest will absorb the equivalent of all the CO_2 in a column of air stacked 1 mile on top of it.[6] Trees aren't merely magnificent vacuum cleaners when it comes to CO_2; they absorb other gases as well: sulfur dioxide generated by coal-burning plants, hydrogen fluoride released by steel and fertilizer plants, and chlorofluorocarbons released by refrigeration and air-conditioning units.[7]

Trees help prevent erosion and purify water. Trees act like gigantic sponges, helping regulate runoff and flooding by gradually absorbing and releasing water. When roots no longer draw water and anchor the soil, when canopies no longer soften the impact of rain and wind, erosion and other consequences are inevitable. Those consequences include everything from tainted water supplies to landslides to the stripping of topsoil from productive land.

Trees help moderate temperatures, in turn reducing fossil fuel consumption and utility company emissions (and your utility bills). Three mature trees shading a house can cut air conditioning costs by up to 50 percent. Planting one hundred million trees—one to two trees per household—could save Americans $4 billion a year on their energy bills.[8] Planted as a windbreak alongside a house, a row of trees can also reduce heating bills in colder climates by up to 20 percent.[9]

Trees are a vital fuel source. Trees are literally a matter of life and death in less developed parts of the world. Wood remains the primary source of heat and cooking fuel for 2 billion people; wood shortages affect about half of these people.[10]

Trees are supreme manufacturing plants. "They make wood in a factory called the forest, using renewable solar energy while absorbing CO_2 in the process," explains Moore.

One source attempts to pin an economic value on the "work" a tree performs in the course of creating wood. The numbers should be taken with a grain of salt, but the study maintains that over its fifty-year life span, one tree generates over $30,000 worth of oxygen, provides $62,000 worth of air pollution control, recycles $37,500 worth of water, and controls $31,250 worth of soil erosion.[11] Another study shows that trees add an average of $9,500 to the resale value of a house.[12] Yet another study shows that hospital patients with a view of trees out their window recovered faster and with fewer complications than those without such views.[13]

Trees are a very good thing. Yet, we have a way of listening to

The "rocket stove" in the background (with the author) is nearly smokeless, burns only small branches, and consumes only a third the amount of wood as the classic three-stone fire in the foreground. Simple devices like this are good for the forests, the environment, and the cook's lungs.

homilies like the above in the same spirit as we do Top 40 songs: we hear them over and over, the melody is familiar, we may even hum along a bit, but we tune out the words. Then one day, we'll actually read the lyrics in the CD liner notes, and suddenly we'll understand what the songwriter has been trying say. "You mean it's not 'All we are is ducks in the wind' but 'dust in the wind'? Oh, that changes things a bit." Sometimes the message is worth listening to; other times, not. This information is worth considering *carefully*.

The relationship between humans, trees, deforestation, and global warming isn't just a complicated issue; it's a *very, very* complicated issue.

For starters, in some countries deforestation isn't an issue per se. Sweden, for example, has twice the volume of standing lumber it did a hundred years ago and today annually "banks" 43 million more cubic yards of wood than it harvests. Canada banks 77 million cubic yards of wood annually.[14] The United States today has about the same amount of forested land it had a hundred years ago. This is due in part to the fact that, even though the population has more than tripled, there's actually slightly less land under cultivation than there was a hundred years ago. And this is partly due to advances in technology, chemistry, biology, and genetics; we can grow five times as much food on the same piece of land as we could a century ago. The relationship between the amount of agricultural land and forested land has always been a tug of war. You can thumb your nose at chemically-goosed crops, but if you love romping in the forest, remember there's a relationship there.

We can decry the deforestation of the Amazon jungle and point fingers, but if your ancestors go back more than a few generations in the United States, chances are they had a hand in the disappearance of 40 percent of America's virgin forests. Clearing forests to create room for farmland and cities, and harvesting trees to create fuel, shelter, and profit, is a pattern that has repeated itself throughout history on a grand scale. [15] In the classical world, unspeakably large tracts of land were cleared to provide fuel for iron, brick, and glass kilns and for the Roman baths. In the Middle Ages in Europe, the population explosion was accompanied by a parallel forest implosion of monumental scale. Forests have always been there for the taking, and humankind has always taken them, often overzealously. But one Brazilian official, in response to all the modern-day finger pointing, said, "If the world wants oxygen, let them pay for it. We're not going to stay poor because the rest of the world wants to breathe."[16]

Deforestation in the tropics is an issue that must be resolved. The situation *does* differ from cataclysmic tree harvests of the past. Much of it is accomplished by, or is accompanied by, slash-and-burn methods that release tremendous amounts of CO_2. The rate is accelerating geometrically in line with geometric population growth. Disproportionate numbers of plant and animal species are disappearing altogether. Rain forest trees, because they grow year round, sequester more CO_2 than do trees in northern forests. Action must be taken—but not with a "you bad, me good" attitude.

If you want further fodder for highlighting the complexity of the situation, chew on this: scientific evidence now suggests that the vast band of boreal forest extending across Canada and Siberia may actually *contribute* to global warming. The dark tree trunks of the dense forests act as a gigantic heat sink that may exacerbate global warming, while a vast barren field of snow in that same area would reflect heat. No one is going to harvest the boreal forest, and certainly there are other tradeoffs and implications, but it does show how counterintuitive the entire situation can be.

We can be dismayed by the felling of trees for human consumption, but again, as Moore points out: when you use less wood, you use more of something else. Until we find that better "something else," trees are our best option. We've yet to find an entity that's as beautiful to behold and, at the same time, as beneficial to mankind as a tree. All in all, trees are a gift. And we should use this gift to our, and our planet's, advantage while making sure there are plenty of gifts to be given in the future.

Many people feel that we need to come down on one side of the fence or the other: are we tree huggers or tree muggers? But the truth is, there should be no fence at all. We need two kinds of forests.

We need protected wilderness areas and national parks where grand trees can grow to a grand old age, where our grandchildren can rest in the shade of the same trees we once did. We should cringe at the thought of hacking down General Sherman to create the three thousand redwood hot tubs it would yield. We should stand in awe of the trees Lincoln, Washington, and Elvis lollygagged under. We should leave the national parks natural, campable, hikeworthy, and forested. A Greek proverb states: "A society grows great when old men plant trees whose shade they know they shall never sit it." Planting a tree is a sign of hope.

On the other hand, we need well-managed forests where we can grow lots and lots of trees. And those trees can provide lots and lots of wood and renewable products, kick out lots and lots of oxygen, and suck in lots and lots of CO_2. Few people protest the harvesting of a corn field; we should have similar fields where we plant, grow, and harvest trees with the same purposefulness.

There's room for both types of forests. We shouldn't get their purposes mixed. Trees in protected forests are something we should relish with glee. And wood from managed forests is something we should consume with gusto. You can disagree with the latter purpose, but as the saying goes, "If you don't think cutting down trees is a good idea, try wiping yourself with plastic toilet paper."

In short (and if you quote me here, be certain to include the *entire* quote), I'm a proponent of deforestation—as long as it's a well-planned deforestation, in the right time and place, followed by an even better planned program of reforestation.

There are gray areas galore, particularly with regard to logging on public lands, but regardless of one's leanings there are things everyone can do.

Plant trees. Plant them in your yard, on your block, and in your parks. Plant them on your own or in conjunction with an organization. American Forests, the National Arbor Day Foundation, and many other organizations have tree-planting programs. (See the Resources section for more information.) In 2007, the United Nations Environmental Program launched an ambitious tree planting campaign called Plant for the Planet: Billion Tree Campaign. Program founder and Nobel laureate Wangari Maathai explained, "Anybody can dig a hole, put a tree in that hole, and make sure that tree survives."[17] What effect will those billion trees have? They'll soak up 250 million tons of carbon dioxide. Impressive.[18]

Buy lumber and other wood-based products that are certified as being grown in sustainably managed forests. There are a number of independently audited systems whereby forests are monitored for sustainablility. These include the Forest Stewardship Council (FSC), the Sustainable Forest Initiative (SFI), the Canadian Standards Association (CSA), and the Pan-European Forest Certification (PEFC).[19] These systems all keep an eye on the big picture; seeing the forest for the trees. They take into account not just the trees but other plants and animals that call the forest home, and those who work in the forests and live near the forest. Certification looks at the soil below, the air above, and the rivers that flow through.[20] Even the big-box stores, Lowes and Home Depot, have policies mandating that FSC-certified wood products be sold or given preference; they monitor suppliers right down to the source of the wood in the carpenter's pencils they sell.

Don't buy CITES-protected wood. The Convention on International Trade in Endangered Species is a treaty, ratified by 160 countries, which regulates the trade of endangered plants (around twenty-eight thousand species) and animals (around five thousand species). Brazilian rosewood is perhaps the best-known wood on the "do not import" list, but lignum vitae, certain mahoganies, sandalwood, and other species are also included.[21] The Rainforest Alliance, through its SmartWood program, certifies over 100 million acres of forest and sixteen hundred wood-related operations in fifty-nine countries. If your wood, wood product, or wood dealer bears the Rainforest Alliance and Forest Stewardship Council (FSC) stamp of approval, you're buying wood from well-managed forests. The process has been monitored from stump to finished product.[22]

Recycle. One ton of recycled paper uses 64 percent less energy and 50 percent less water, generates 74 percent less air pollution, saves seventeen trees, and creates five times more jobs than one ton of paper produced from virgin pulpwood. If every person in America recycled their Sunday paper every week, we'd save twenty-six million trees per year.[23]

Use technology to make things better. We should accelerate the pace at which we take the technologies used in developed countries to quintuple crop yields and make them readily available to countries with vanishing forests. This step makes low-yielding farmland created through slash-and-burn methods more productive, and it helps break the relentless cycle of deforestation required to clear more and more low-yielding land.

In a similar vein, making wood preservative technology readily available to developing countries would have a beneficial effect. Wood is often shunned as the primary building material in tropical areas because of termites and rot. Concrete and other less environmentally friendly materials are often used instead. And in countries where people use more wood for housing and other purposes, people tend to plant more trees.

Can one tree provide $62,000 worth of air pollution control, add $9,500 to the resale value of your house, cut your air conditioning costs in half, and help you recover more quickly when you're in the hospital? Maybe.

In the end, what to do is a personal decision. But in this topsy-turvy world, where it can be difficult to make a tangible difference, being able to point to a tree you've planted—one that generates oxygen, sequesters CO_2, cools your house, calms an ailing heart, provides woods for your grandchildren to play in, and supplies wood for your great-grandchildren to build with—is heartening.

John Muir wrote, "God has cared for these trees, saved them from drought, disease, avalanches, and a thousand tempests and floods. But he cannot save them from fools." There are two kinds of fools: those who waste trees and those who do not use them to full advantage. Let's not be fools on either count.

NOTES

CHAPTER 1

1. Lake, "The Magic of Huon Pine," 30. *Note:* Almost as rare as buried ancient Huon pine is standing, living Huon pine; only 1,050 trees remain. And in the world of rare lumber supply and demand, a wood stump containing 1 cubic meter of birdseye Huon pine was recently purchased by a Japanese veneer maker for $35,000.

2. Lewington and Parker, *Ancient Trees: Trees That Live for a Thousand Years,* 136. *Note:* By 1920, that 4 million acres of kauri forest had been reduced to 18,420 acres; 0.5 percent of the original virgin forest.

3. Lewington and Parker, 139.

4. Information in this section was obtained through telephone interviews with Mitch Talcove, Sam Talarico, and Rick Hearne in 2006. See Resources for more information.

5. Logan, *Oak: The Frame of Civilization*, 97.

6. DeWitt, "A Favorite Wood."

7. Logan, 37.

8. Logan, 55.

9. Portugese Cork Association, www.realcork.org.

10. Jelinek Cork, http://www.jelinek.com/about_cork.htm

11. http://www.morlanwoodgifts.com/MM011.ASP?pageno=149.

12. Telephone interview with Patrick Moore, May 25, 2007.

13. Constantine, *Know Your Woods,* 68.

14. Snyder, "How Is Birdseye Maple Formed?"

15. Lacer, "Spalted Wood."

16. www.ag.auburn.edu/aaes/communications/bulletins/figureinwood/index.

17. Beals and Davis, "Figure in Wood: An Illustrated Review."

18. Beals and Davis.

19. http://www.woodworkforums.ubeaut.com.au/archive/index.php/t-34190.html.

20. Constantine, *Know Your Woods,* 297.

21. Porter, *Wood: Identification and Use,* 27–32.

22. For more specific information regarding wood figure and wood density, see Hoadley, *Understanding Wood,* and Constantine, *Know Your Woods.*

23. Wong, "High-Rise Bamboo Scaffolds Can Make for a Deadly Climb.

24. Much of the information in this section comes from the author's interview with Jeff McMullin of McMullin Sawmill in June 2006.

25. Preston, "Climbing the Redwoods."

26. Preston.

27. Much of the information in this section comes from an interview with Liz Bieter, formerly of the Duluth Timber Company in June 2006.

28. "The Lumberyard" on *Modern Marvels;* The History Channel; aired March 27, 2007.

29. Dimech, The Plant Evolution Tour (parts 1 through 15), http://adonline.id.au/plantevol/tour/.

30. Farb, *The Forest,* 39–71.

31. Russell, *The New Encyclopedia of American Trees,* 10–11. *Note:* The exact number of tree species is open to debate. Those supporting the definition of a tree as a plant that's long-lived and 10 centimeters or more in diameter (in addition to having a single woody stem) can put the number as low as twenty-five thousand.

32. Morlan, Facts About Wood & Trees, 2. http://www.morlanwoodgifts.com/. *Note:* Again, depending on one's definition of "tree," some put the number of tree species in the United States at far fewer than 1,182.

33. Farb, 57–72.

34. Hoadley, *Understanding Wood,* 7.

35. E-mail from Regis Miller to the author, November 11, 2007.

36. Aidan Walker, *The Encyclopedia of Wood,* 8–15.

37. Russell, *The New Encyclopedia of American Trees,* 14–17.

38. Harlow, *Inside Wood: Masterpiece of Nature.* Harlow is also a good reference for information on wood and tree anatomy.

CHAPTER 2

1. Groeschen. *The Art of Chainsaw Carving,* 113.

2. Based on author's visit with Livio De Marchi, April 18, 2006, Venice, Italy.

3. Fusco, "Blind Woodworker Crafts Lifelong Dream."

4. John Cook, "Blind Woodworker's Story," www.woodweb.com, January 21, 2003.

5. Parts of this section are based on e-mails from Larry Martin and phone interviews with Gordon Mitchell, David Albrektson, and Ron Faulkner in December 2006.

6. Gordon Mitchell, "In the Workshop Without Vision." WoodCentral Web site, www.woodcentral.com.

7. "Vision for a Challenge: Gordon Mitchell," *Woodcraft,* as found at www.woodcraftmagazine.com.

8. *Guinness World Records 2006,* 64.

9. E-mail from Norm Satorius, March 25, 2007.

10. Note the difference between "specimens" or "samples" and "species." There may be many specimens or samples—a cross-section of a branch, a slice of the trunk with bark intact, samplings of the heartwood and sapwood—from a single species in a collection.

11. Miller. *Xylaria at the Forest Products Laboratory,* 1.

12. Much of the information in this segment is based on the author's interview with Mira Nakashima, New Hope, Pennsylvania, on January 23, 2007.

13. Mira Nakashima, *Nature, Form and Spirit,* 25.

14. George Nakashima, *The Soul of a Tree,* 93.

15. George Nakashima, 128.

16. George Nakashima, 116.

17. Mira Nakashima, 9.

18. Mira Nakashima, 264.

19. The bulk of the information in this segment is based on the author's telephone interview with President Carter, March 6, 2007.

20. Carter, *An Hour Before Daylight,* 36.

21. Pacher, "The Restoration of Jimmy Carter."

22. Carter, *Sharing Good Times,* 158.

23. Habitat for Humanity Web site: http://www.habitat.org/how/carter.aspx.

24. Carter, *Sharing Good Times,* 156.

25. Pacher, "The Restoration of Jimmy Carter."

CHAPTER 3

1. Rybczynski, *One Good Turn,* 89.

2. The bulk of the information in this section is based on a series of interviews conducted by the author with Alan Lacer in 2006.

3. *The Gristmill,* December 2006, What's It, 29.

4. Powell, letter.

5. Barker, 4.

6. Barker, 7e.

7. Barker, 18

8. Staten, 144.

9. Most information in this section is from the Power Tool Drag Races Web site, http://powertooldragraces.com/rules.html.

10. Much of the information in this story came from the author's participation in the Sixth Annual Tyrol Basin Belt Sander Races held on March 11, 2007, near Mount Horeb, Wisconsin. For information on future races visit www.tyrolbasin.com.

CHAPTER 4

1. Christie's Auction, May 16, 2006.

2. Much of this information is based on interviews conducted at the Cremona International School for Violin Making, April 27, 2006, Cremona, Italy.

3. From literature from SoundWood, an organization dedicated to ensuring the long-term supply of specialty woods used in musical instrument making, www.soundwood@fauna-flora.org. Grauer, February 27, 2003.

4. Graver. *Cigar Aficianado* online article, winter 1995.

5. Bonornetti, 135–141.

6. Michael D. Lemonick, *Discover*, July 2000.

7. Letter to Signor Mandelli from Giacomo Stradivari, approximately 1860.

8. Reid, 10.

9. Choi.

10. "Stradivari," *60 Minutes*.

11. Information based on interviews with James Olson of Olson Guitars, Circle Pines, Minnesota, in March 2006.

12. Andrews, 43.

13. "History of Drums," http://penz4.tripod.com/historyofdrums.html.

14. Monteux.

15. Barron, 89.

16. Warren Albrecht, e-mail, February 6, 2007.

17. Chapin, 87–89.

18. Information based on author's interview with John Koster at National Music Museum, September 2006.

CHAPTER 5

1. Gutman, 30.

2. Hill, 108.

3. Hill, 10–12.

4. Gutman, 2.

5. Gutman, 12.

6. King, 16.

7. Pat Ryan, "Golf: A History of Golf Clubs," http://www.patryangolf.com/2006/pat_ryan_golf_other_interesting_stuff/pat_ryan_golf_club_history.htm.

8. Betts.

9. Potter.

10. Yoder.

11. "Wooden Soldiers," 23.

12. Elmore Just, "The Persimmon Story," http://www.louisvillegolf.com/story.php.

13. Celebration of Celts: Festival of Music, Dance and History Web site, http://www.celebrationofcelts.com/athletics.html.

14. http://www.electricscotland.com. Click on games and then Caber Toss.

15. National Center for Catastrophic Sports Injury Research, www.unc.edu/depts/nccsi/AllSport.htm.

16. Boomerang Association of Australia Web site, http://www.boomerang.org.au/articles/article-what-is-a-boomerang.html.

17. Fisher.

18. http://www.itftennis.com/technical/equipment/rackets/history.asp.

19. Fisher.

20. Joe Sch artman, www.WoodTennis.com, e-mail interview, April 10, 2006.

21. Mark McClusky, "Tennis Swaps Grace for Strength," Wired News, August 26, 2003, http://www.wired.com/news/technology/1,60177-1.html.

22. The bulk of this information came from interviews conducted during the Eleventh Annual Midwestern Lumberjack Championships held during the Rochesterfest 2006, Summer Daze lumberjack competition, June 17–18, 2006, Foster Arend Park, Rochester, Minnesota.

CHAPTER 6

1. Nelson et al., 42.

2. Parkenham, 138.

3. Ashford, 30.

4. Out 'n' About Treesort Web site, www.treehouses.com.

5. Ashford, 33.

6. Ron Emmons, "The Bizarre Baobab," ronemmons.com/nature/baobabs/.

7. McRaven, 2.

8. Johnson, 22.

9. Johnson, 22–24.

10. Frazer, Chapter 9, section 1, http://www.sacred-texts.com/pag/frazer/.

11. Frazer, ibid.

12. O'Brien.

13. Western Wood Products Association. *Product Use Manual*, 7–9.

14. Hoadley, *Understanding Wood*, 80.

15. Western Wood Products Association, *Product Use Manual*, 15.

16. Salvadori, 48.

17. Levy and Salvadori, 56.

18. Levy and Salvadori, 29

19. "Engineered Wood and the Environment."

20. Halvorsonetal.,20.www.knightridder.com/about/greatstories/charlotte/house28.html.

21. Halvorson et al.

22. Halvorson et al.

23. American Lung Association, www.healthhouse.org/iaq/buildingscience.asp.

24. "Canine Mold Detectives and Infrared Cameras: A New and Better Way to Wage 'War on Spores,'" http://www.prweb.com/releases/2004/6/prweb134304.htm.

25. Halvorson et al.

26. Much of the information in this segment came from Troy Taylor. "The Winchester Mystery House."

CHAPTER 7

1. Silverstein. *The Giving Tree.*

2. Peattie. *A Natural History of Trees,* 456.

3. Auden, 145–146.

4. *International Book of Wood*, p. 9.

5. Perlin, 168.

6. Perlin, 125.

7. Hair; Hall and Maxwell.

As Peattie states in *A Natural History of North American Trees* (34): Part of America's vast lumber consumption consisted of waste. In twenty-four years the states of Michigan, Wisconsin, and Minnesota alone produced 85 billion pine shingles. Only the choicest parts of the tree were used. "The sapwood, the knots, much of the heart and practically the whole trunk above the first 20 feet were left in the woods to rot. It was not unusual to sacrifice a 3000-[board]foot tree to get 1000 shingles—throwing away about 14/15ths and using 1/15th."

8. Leslie, 328–338.

9. Based on author's visit to the Mercer Museum, January 23, 2007.

10. Moore, 93.

Statistics on wood consumption are all over the map and in some cases vary from one source to another—even reliable sources—by as much as 400 percent.

11. Moore, 93.

12. http://www.idahoforests.org/wood_you.htm.

13. http://www.sciencecases.org/taxol/taxol.asp.

14. http://www.mda.state.mn.us/MAITC/forestry.pdf.

15. Youngs, 216.

16. Christiansen, 8.

17. Fisher, *The Lindbergh Case.*

18. Christiansen, 9.

19. Batcha.

20. Adkins, 44.

21. Mary Bellis, "A Brief History of Writing Instruments," http://inventors.about.com/library/weekly/aa100197.htm?once=true&.

22. Evan Lindquist, "Recipes for Old Writing and Drawing Inks," http://www.clt.astate.edu/elind/oldinkrecipes.htm.

23. Petroski, 56.

24. Peattie, *A Natural History of North American Trees*, 133.

25. Petroski, 202–204.

26. http://www.pencilpages.com/articles/zora-art.htm.

27. Petroski, 6.

28. Kilby, 28.

29. "World's Largest Wooden Barrel Factory Flourishes in Missouri."

30. http://www.cellarnotes.net/wine_barrels.htm.

31. http://www.beekmanwine.com/prevtopah.htm.

32. http://www.beekmanwine.com/prevtopah.htm.

33. http://www.foodreference.com/html/artbarrels.html.

34. "Inside the Hoops: A Guide to Better Cooperage."

35. Logan, 179.

36. I Kings 18:33.

37. Logan, 180.

38. "The True Cross," http://www.newadvent.org/cathen/04529a.htm.

39. Thiede and d'Ancona, 3.

40. "The True Cross, Growth of the Christian Cult," http://www.newadvent.org/cathen/04529a.htm.

41. Thiede and d'Ancona, 54.

42. Weisse.

43. Lehane, 235.

44. Thiede and d'Ancona, 9.

45. Plotnik, 270.

46. Constantine, 77–78.

47. http://en.wikipedia.org/wiki/Spear_of_Destiny. Other true relics Web sites: "Relics attributed to Jesus," http://en.wikipedia.org/wiki/Category:Relics_attributed_to_Jesus. "The Discovery of the True Cross of DNJC," http://www.ichrusa.com/courtyard/skulljohnbaptist.html. "The True Cross," http://www.newadvent.org/cathen/04529a.htm.

48. Hubbell, 76–78.

49. "Mouthpiece: The Newsletter of Ashwater Dental Practice," http://www.ashwaterdental.co.uk/Scrapbook/mouthpiecespring99.htm.

50. Green, *Wood: Craft, Culture, History*, 277.

51. http://www.toiletpaperworld.com/tpw/encyclopedia/navigation/funfacts.htm.

52. www.foodreference.com/html/ftoothpicks.html.

53. E-mail interview with Stan Munro, August 1, 2006; telephone interview in August.

CHAPTER 8

1. Payne-Gallwey, 33.

2. Payne-Gallwey, 29.

3. Gurstelle, 114.

4. Based on author's telephone interview with William Gurstelle, February 14, 2007.

5. Payne-Gallwey, Part 1, Introductory Notes.

6. "Leonardo's Dream Machines."

7. "Ancient Discoveries—Machines I and II."

8. "Ancient Discoveries—Machines I and II."

9. Volkman, 1.

10. "The Hundred Years War," www.hyw.com.

11. Alex Song (*Thorn* columnist), *The Rose Thorn*, CM 5037, 5500 Wabash Ave., Terre Haute, IN 47803. "Buick to face medieval wrath" online article at www.rose-hulman.edu/thorn/archive/index.php?issue=980904.

12. Based on author's telephone interview with William Gurstelle, February 14, 2007.

13. "USS *Constitution* Timeline" (compiled by Capt. Steven Maffeo, USNR), http://www.ussconstitution.navy.mil/historyupdat.htm#1794-1797.

14. Dewitt, 21.

15. "Southern Comfort," http://www.americanforests.org/productsandpubs/magazine/archives/2006winter/feature1_1.php.

16. Ibid.

17. Barkman.

18. www.vasamuseet.se.

19. McNeil, online article, http://query.nytimes.com/gst/fullpage.html?res=9D02E6DB1439F937A25756C0A9649C8B63.

20. www.vasamuseet.se.

21. "International Evaluation of the Preserve the *Vasa* Project," 18, http://www.vasamuseet.se/upload/bevara_vasa_rapport.pdf.

22. Volkman, 46.

23. Grace.

24. Hardy, 130.

25. Hardy, 12.

26. Keegan, 119.

27. Hurley, 217.

28. Hurley, 13.

29. Hurley, 15; Clarence Ellsworth, "Bows and Arrows" (leaflet) in the Southwest Museum.

30. "Long Bow Making," www.naturalbows.com and www.archery.com.

31. Lewington Parker, 71.

32. Hurley, 9, 216.

33. *International Book of Wood,* 210.

34. Volkman, 51.

35. "An Introduction to Archery," http://www.yale.edu/archery/about.html.

36. Much of the information in this segment comes from the author's visit to Thomas Boehm's Ancient Archery workshop and facility, in December 2006.

37. Perlin, 281.

38. Beresford-Kroeger, 105.

39. Peattie, *A Natural History of Trees,* 4.

40. Perlin, 280.

41. Heinrich, 213.

42. Peattie, *A Natural History of North American Trees,* 30.

43. Peattie, *A Natural History of North American Trees,* 8.

44. Perlin, 296.

45. Peattie, *A Natural History of North American Trees,* 32.

46. Perlin, 294.

47. Independence Hall Association of Philadelphia, http://www.ushistory.org/betsy/flagpics.html.

48. Tsutsui.

49. "US Naval Technical Mission to Japan: Reports in the Navy Department Library." http://www.history.navy.mil/library/guides/japan.htm.

50. Tsutsui.

51. Jeffries.

52. "Joint DOE, USDA Grant Funds Development of Poplar Trees Optimized for Ethanol Feedstock," October 26, 2006, www.greencarcongress.com.

53. Tuskan.

54. Oo and Falk.

CHAPTER 9

1. Royal Navy Museum.

2. *International Book of Wood*, 158–159.

3. Porter, 141.

4. Barton, 83.

5. Barton, 85.

6. Huit, one-page printout from Evergreen Museum.

7. Schwartz, 22.

8. Barton, 83.

9. Huit, one-page printout from Evergreen Museum.

10. Barton, 85.

11. Barton, 87.

12. Scott.

13. Patent application No. 821,393 for O. & W. Wright Flying Machine, filed March 23, 1903.

14. http://www.sensenichprop.com.

15. *Sir George Cayley (1773–1857)*." London: Gibbs-Smith, 1968, 21, as quoted Pelham, 26.

16. http://www.nationalkitemonth.org/history/kitehistory.shtml.

17. Pelham, 28.

18. Pelham, 24.

19. Pelham, 37.

20. Pelham, 45.

21. Pelham, 72.

22. White.

23. Pelham, 30.

24. Hart, 147.

25. Hart, 150.

26. "Did You Know and Kite Facts," www.kiteman.co.uk.

27. Hart, 25, 159.

28. "Did You Know and Kite Facts," www.kiteman.co.uk.

29. Perlin, 229–231.

30. Nowlin, 192.

31. Freese, 122.

32. John White, "Wood to Burn," http://www.americanheritage.com/articles/magazine/ah/1974/1/1974_1_78.shtml.

33. Ibid.

34. Freese, 122.

35. Freese, 124.

36. "Reform in Manhattan," www.time.com/time/archive/preview/0,10987,776133,00.html.

37. "Everlasting Steam: The Fate of Jupiter and 119," http://www.nps.gov/gosp/history/everlasting_steam.html.

38. White, "Wood to Burn."

39. "Recycled Plastic Lumber," http://www.plasticsresource.com/s_plastics resource/sec.asp?TRACKID=&CID=128&DID=230.

40. http://www.rta.org/.

41. *The New Oxford Annotated Bible,* Genesis 6:15.

42. "Biblical Mount Sinai," http://en.wikipedia.org/wiki/Biblical_Mount_Sinai.

43. Richie.

44. "What is Gopher Wood?" http://christiananswers.net/q-eden/gopher wood.html.

45. "Doctoral Student Weighs the Cost, Structure of a Famous Ship," www.whistle.gatech.edu/archives/04/apr/19/ark.shtml.

46. Association El Felze, Venice, Italy, www.elfelze.com/english-gondola.html.

47. http://www.squero.com/.

48. Donatelli, 110.

49. Much of the information in this segment is based on the author's interview with gondolier John Kirschbaum, January 15, 2007.

50. www.tramontingondole.it.

51. http://www.elfelze.com/english-elfelze.html.

52. Giorgi, "Il lavoro del paio."

53. Vittoria, 167.

54. Vittoria, 168.

55. Association El Felze, http://www.elfelze.com/english-elfelze.html.

56. Giorgi, "Forcole E Remi, Squerie Paline," 35.

57. Association El Felze, http://www.elfelze.com/english-remeri.html.

58. Giorgi, "Forcole E Remi, Squerie Paline," 35.

CHAPTER 10

1. "Peace in Space," http://www.theguardians.com/space/orbitalmech/gm_sspow.htm.

2. Giorgi, "Il lavoro del paio."

3. Hoadley, *Understanding Wood,* 80.

4. "UK Timber for Marine and Geotechnical Applications," April 13, 2004, http://www.forestry.gov.uk/pdf/crwoodproducts26.pdf/$file/crwoodproducts 26.pdf.

5. Brown.

6. "Gorski kota," http://en.wikipedia.org/wiki/Gorski_kotar.

7. Walker, 121.

8. Lauritzen, 42.

9. http://www.timberpilingcouncil.org/history.html.

10. "Foundation Piling: How Long Will It Last?" www.zeta.org.au.

11. Pollio, *The Ten Books of Architecture,* chapter 9, paragraph 10, online at www.penelope.uchicago.edu/Thayer/E/Roman/Texts/Vitruvius/home.html.

12. Canadian Wood Council, http://www.cwc.ca/publications/brochures/green_by_design/energy.php.

13. "UK Timber for Marine and Geotechnical Applications," April 13, 2004, http://www.forestry.gov.uk/pdf/crwoodproducts26.pdf/$file/crwoodpro ducts26.pdf.

14. UNESCO, http://portal.unesco.org/culture/fr/ev.php-URL_ID=3506&URL_DO=DO_TOPIC&URL_SECTION=201.html.

15. Much of the information in this section is from the author's interviews with Sebastiano Giorgio and Elena Barinova, April 17, 2006, Venice, Italy.

16. "Wooden Water Pipe."

17. *International Book of Wood,* 150.

18. Hope.

19. http://sewerhistory.org/images/pi/1924_pi02.jpg.

20. "Wooden Water Pipe."

21. http://www.sewerhistory.org/grfx/components/pipe-wood3.htm

22. Much of the information in this section came from the author's interview with Jon Schladweiler, January 10, 2007.

23. Cook, 7.

24. Brokaw, 14.

25. Easley.

26. Ferruci and Acevedo-Rodriquez, 153–162.

27. Davidson.

28. Adkins, 67.

29. http://ias.berkeley.edu/orias/visuals/japan_visuals/shintoC.HTM.

30. Andre Alexander, "The Lhasa Jokhang—Is the World's Oldest Timber Building in Tibet?" http://www.webjournal.unior.it/Dati/17/47/Articolo%208%20Alexander.pdf.

31. "World's Oldest Wooden Pagoda in Badly Need of Repair," People's Daily Online. http://english.people.com.cn/200609/01/eng20060901_298699.html.

32. "World's Oldest Building Discovered," BBC News, March 1, 2000, http://news.bbc.co.uk/1/hi/sci/tech/662794.stm.

33. Tillamook Air Museum, http://www.tillamookair.com/html/bldg.html.

34. "The Virginian Pilot," August 4, 1995, http://scholar.lib.vt.edu/VA-news/VA-Pilot/issues/1995/vp950804/08040464.htm.

35. Lane Memorial Library, http://www.hampton.lib.nh.us/hampton/history/pamphlets/bridgestatistics.htm.

36. "History: Past, Present and Future," http://www.trestlewood.com/about_tw/history/history.html.

37. Frederick Huchel, "History of Box Elder County," http://historytogo.utah.gov/utah_chapters/statehood_and_the_progressive_era/thelucincutoff.html.

38. www.trestlewood.com.

39. Much of the information in this segment is based on the author's interview with Ken Felber, construction manager, Valley Fair, Shakopee, Minnesota, January 12, 2007.

40. http://www.ultimaterollercoaster.com/coasters/history/1980_1990/90s_wood.shtml.

41. Scott Rutherford. *The American Roller Coaster,* 13.

42. Goldsmith.

43. http://www.enquirer.com/editions/2000/04/05/loc_getting_son_of_beast.html.

44. E-mail interview with Richard Bannister, August 7, 2006.

45. http://www.ultimaterollercoaster.com/forums/cgi/forum1.cgi?read=153293 (posted by "frontrunner").

EPILOGUE

1. "Exploring the Environment," http://www.cet.edu/ete/modules/temprain/trwood4.html.

2. National Arbor Day Foundation, www.arborday.org.

3. Earth Policy Institute, http://www.earth-policy.org/Indicators/Forest/2006.htm.

4. Information and statistics about the benefits of trees vary greatly and are open to varying degrees of interpretation.

5. Cooper.

6. Phone interview with Patrick Moore, May 27, 2007.

7. Russell, 30.

8. Little, 205.

9. Minnesota Department of Natural Resources Web site, "Reasons to plant trees," http://www.dnr.state.mn.us/forestry/nurseries/reasons.html.

10. Perlin, 15. According to a study by Deon Filmer and Lant Pritchett, "Environmental Degradation and the Demand for Children: Searching for the Vicious Circle," www.worldbank.org/html/prddr/prdhome/peg/wps02/indexp2.htm, in rural Pakistan, where 80 percent of households use firewood for cooking, anywhere from 700 to 1,050 "people hours" per year are spent collecting firewood. The average distance traveled is about 1 mile each way, requiring about 2 hours per day.

11. USDA Forest Service Pamphlet #R1-92-100.

12. David J. Nowak, "Benefits of Community Trees," (Brooklyn Trees, USDA Forest Service General Technical Report, in review).

13. USDA Forest Service Pamphlet #R1-92-100.

14. Porter, 11.

15. Moore, 92.

16. Williams, 476.

17. "A Billion Trees?" 20.

18. United Nations Environment Programme, "Fast Facts," www.unep.org/billiontreecampaign/FactsFigures/FastFacts/index.asp.

19. E-mail interview with Patrick Moore of Greenspirit, January 21, 2008.

20. For more information and a searchable database of companies following FSC guidelines, visit www.fscus.org or www.fsc.org.

21. CITES, Appendices I, II, and III, http://www.cites.org/eng/app/appendices.shtml.

22. The Rainforest Alliance Web site also includes a complete list of certified companies and products at http://www.rainforest-alliance.org/forestry/documents/smartguide_construction.pdf.

23. "Recycling Facts," University of Colorado at Boulder, http://recycling.colorado.edu/education_and_outreach/recycling_facts.html.

RESOURCES

WOODS

Ancientwood, Ltd.
Ashland, WI
www.ancientwood.com
North American supplier of ancient kauri wood

Duluth Timber Company
Duluth, MN
www.duluthtimber.com
Timbers, flooring, and millwork from reclaimed old-growth timbers

Hearne Hardwoods
Oxford, PA
www.hearnehardwoods.com
Over a hundred species of domestic and exotic hardwood lumber in stock

McMullin Sawmill
Crescent City, CA
www.sawemup.com
Custom-sawn old-growth redwood from snags, windfalls, and recycled timber

Talarico Hardwoods
Mohnton, PA
www.talaricohardwoods.com
World-class, highly figured, and book-matched lumber

Timeless Timber
Ashland, WI
www.timelesstimber.com
Old-growth timber recovered from North American lakes and rivers

Tropical Exotic Hardwoods of Latin America, LLC
Carlsbad, CA
www.anexotichardwood.com
Exotic tropical lumber, including cocobolo, amboyna burl, and other woods

WOODWORKERS, WOOD PRODUCT MANUFACTURERS, TOOLMAKERS, ETC.

Ancient Archery
Dodgeville, WI
www.ancientarchery.com
Handcrafted longbows, flat bows, and recurve bows; arrows

Arnot Q Custom Cues
Lake Fort, FL
www.arnotq.com
High-end custom pool cues, pool cue kits, cue-making school

The Barrel Mill
Avon, MN
www.thebarrelmill.com
Cooperage specializing in premium wine and whiskey barrels

Livio De Marchi
Venice, Italy
www.liviodemarchi.com
Wood carver extraordinaire, specializing in replicated clothes, cars, and other objects

George Nakashima Woodworker, S.A.
New Hope, PA
www.nakashimawoodworker.com
Freeform tables, chairs, furniture, and cabinets built in the spirit of George and Mira Nakashima; oversees Nakashima Peace Foundation

Gondola Romantica
Stillwater, MN
www.gondolaromantica.com
Authentic gondola rides on the scenic St. Croix River

James A. Olson Guitars
Circle Pines, MN
www.olsonguitars.com
Custom-made acoustic guitars, including those for James Taylor

Alan Lacer, Woodturner
River Falls, WI
www.alanlacer.com
Videos, woodworking tools, and classes in the fine art of wood turning

Louisville Golf
Louisville, KY
www.louisvillegolf.com
Wood drivers, fairway woods, and putters made of persimmon, walnut, cherry, maple, pear, and other fine woods

MaxBats
Brooten, MN
www.maxbats.com
Custom-made maple baseball bats

Johannes Michelsen
Manchester Center, VT
www.woodhat.com
Custom-fitted, turned wood cowboy, hats, bowlers, baseball caps, and other "head wear"; also classes and tools

Stan Munro
North Syracuse, NY
www.toothpickcity.com
Creator of Toothpick City

Power Tool Drag Races
www.powertooldragraces.com
Annual power tool drag races with a variety of classes, including super stock, pro-supercharged, awful altereds, ridden, and others

SawStop, LLC
Wilsonville, OR
www.SawStop.com
Manufacturer of SawStop table saws

Sculpture by Roghair
Hinckley, MN
www.sculpturebyroghair.com
Chainsaw sculptures, including animals, people, and landmarks, by Dennis Roghair

Steinway & Sons
Long Island City, NY
www.steinway.com
Grand and upright pianos in a wide variety of models; showrooms in New York City, Miami, and other locations

Double Diamond Belt Sander Races
Tyrol Basin
Mt. Horeb, WI
www.tyrolbasin.com
Annual belt sander races with open and modified classes; competitions are normally held in March

The Wild Mountain Man Ray Murphy
Hancock, ME
www.thewildmountainman.com
Murphy—the world's first chainsaw artist—has been creating chainsaw art since 1957, using chainsaws, period, when working; saw-sculpting shows and art gallery

SCHOOLS, ORGANIZATIONS, AND MUSEUMS

American Association of Woodturners
St. Paul, MN
www.woodturner.org
Publishes *American Woodturner* magazine quarterly; sponsors shows, symposiums, meetings; 275 North American chapters

The Carter Center
Atlanta, Georgia
www.cartercenter.org
The Center, in partnership with Emory University, focuses on alleviating human suffering, preventing and resolving conflicts, enhancing freedom and democracy, and improving health

Evergreen Aviation Museum
McMinnville, OR
www.sprucegoose.org
Home of the *Spruce Goose* as well as fighter, bomber, passenger, home-built, and other planes

Forest Products Laboratory
Madison, WI
www.fpl.fs.fed.us
250 scientists and support staff who conduct research on diverse aspects of wood use; wood identification information and analysis

Forest Products Society
Madison, WI
www.forestprod.org
Provides a variety of publications and conference opportunities for companies and individuals using wood and wood-fiber resources

Sebastiano Giorgi
Venice, Italy
Journalist and publisher affiliated with *Lagunamare* bimonthly review; www.assonauticavenezia.it and www.iantichieditori.it

Habitat for Humanity International
Americus, GA
www.habitat.org
Seeks to eliminate poverty housing and homelessness from the world and to make decent shelter a matter of conscience and action; has built over two hundred thousand dwellings in three thousand communities for over one million people; the Jimmy Carter Work Project is an annual weeklong housing blitz where thousands of volunteers build homes, alternating between locations in this country and those abroad

International School of Violinmaking
Cremona, Italy
www.scuoladiliuteria.com
Three- and five-year programs in stringed instrument making

International Wood Collectors Association
www.woodcollectors.org
For those interested in collecting and identifying wood and using wood in creative crafts; $35 membership fee includes *World of Wood* magazine

Loretto Chapel Miraculous Staircase
Santa Fe, NM
www.lorettochapel.com
Freestanding, wood spiral staircase that makes two 360-degree turns; one of the Seven Wonders of New Mexico; open for tours year round

Mid-West Tool Collectors Association
www.mwtca.org
International organization for studying, preserving, and understanding early tools and devices; publishes *The Gristmill* quarterly

National Music Museum
The University of South Dakota
Vermillion, SD
www.usd.edu/smm
Over thirteen thousand instruments from around the world; specialized collections of Conn instruments, harpsichords, early Italian stringed instruments, and harmonicas; also one of the most fabulous guitar collections in the world

Sewerhistory.org
www.sewerhistory.org
Web site dedicated to displaying Jon Schladweiler's collection of sewage conveyage systems, spanning 3500 BC through the 1930s

SoundWood, Fauna & Flora International
www.soundwood.org
Dedicated to the preservation and sustainability of "tonewoods" through education, habitat conservation, and wood certification

USS *Constitution*
Charlestown, MA
www.ussconstitution.navy.mil
Naval museum and home of *Old Ironsides*; daily tours and special events

The *Vasa* Museum
Stockholm, Sweden
www.vasamuseet.se
Home of the restored *Vasa* warship and other nautical artifacts

Warwick Castle
Warwickshire, England
www.warwick-castle.com
"Britain's Greatest Mediaeval Experience"

Winchester Mystery House
San Jose, CA
www.winchestermysteryhouse.com
Eccentric Victorian mansion built by Sarah Winchester over the course of thirty-eight years, containing 160 rooms, 47 fireplaces, 467 doors, and other oddities; daily tours

Woodworking for the Blind
Deer Park, IL
www.woodworkingfortheblind.org
Provides free voice recordings of woodworking publications for use by blind and visually impaired woodworkers; publications recorded include *Fine Woodworking, Woodwork, Woodsmith, Woodworking,* and *American Woodworker*

World Forestry Center
Portland, OR
www.worldforestrycenter.org
Nonprofit educational organization dedicated to informing people about the importance of world forests and trees; includes discovery museum, tree farms, and World Forest Institute

TREE PLANTING INITIATIVES, PROGRAMS, AND SUPPORTERS

American Forests
Washington, DC
www.americanforests.org
Nonprofit organization focused on fostering healthy forest ecosystems; Global ReLeaf program has planted over twenty-three million trees; publishes *American Forests* quarterly magazine and *National Register of Big Trees*

Convention of International Trade in Endangered Species of Wild Fauna and Flora (CITES)
www.cites.org
International agreement which strives to ensure that international trade in animal and plant specimens does not threaten their survival

Forest Stewardship Council
www.fsc.org
International organization that sets standards for responsible forest management, certification of forest management programs, information, and education

Greenspirit: For a Sustainable Future
www.greenspirit.com
Web site by Patrick Moore, author of *Green Spirit: Trees Are the Answer,* which promotes ideas for a sustainable future

A variety of companies offer tree-related initiatives and incentives to those purchasing or using their products. Companies include Baby Appleseed, www.babyappleseed.com; Computershare "eTree" program, www.etreeusa.com; Enterprise Rent-A-Car, www.arborday.org/enterprise/intro.cfm; IKEA, www.ikea.com/ms/en_US/about_ikea/plant_trees.html; and Smith Barney, www.smithbarney.com/trees

Contact your state Department of Natural Resources to see if they have a seedling program. One example is the Minnesota program, where trees must be ordered in minimum lots of five hundred (and be planted in Minnesota). Prices are reasonable: One thousand Norway pine seedlings (6 inches to 12 inches tall) cost $170. Visit http://www.arborday.org/programs/urbanforesters.cfm for a state-by-state listing of similar forestry programs.

The National Arbor Day Foundation
www.arborday.org
Rain Forest Rescue Program, "Trees to the Forest" reforestation program, Trees for Katrina, "Give-a-tree" gift cards, "Trees in Celebration," and "Trees in Memory" programs

TreeLink
www.treelink.org
A clearinghouse for information on urban and community forests and tree care

United Nations Environment Program
www.unep.org
Plant for the Planet: Billion Tree Campaign along with lots of literature, programs, and information regarding trees

BIBLIOGRAPHY

BOOKS

Adkins, Jan. *The Wood Book*. Boston: Little, Brown and Company, 1980.

Andrews, George, ed. *The American History and Encyclopedia of Music*. London: Irving Square, 1908.

Arbor, Marilyn. *Tools and Trades of America's Past: The Mercer Museum Collection*. Doylestown, PA: Tower Hill Press, 1981.

Auden, W. H. *Selected Poems of W. H. Auden*. New York: The Modern Library, 146.

Barker, Harold. "A Pictorial History of the American Circular or Table Saw (1800–1960)." Ada, OH: Compilation of photocopied articles (undated).

Barron, James. *Piano: The Making of a Steinway Concert Grand*. New York: Times Books, 2006.

Barton, Charles. *Howard Hughes and His Flying Boat*. Blue Ridge Summit, PA: AERO Books, 1982.

Beresford-Kroeger, Diana. *Arboretum America: A Philosophy of the Forest*. Ann Arbor: University of Michigan Press, 2003.

Brokaw, Bill. *Loretto Chapel: The Mysterious Staircase*. Lawrenceburg, IN: The Creative Company, 2002.

Carter, Jimmy. *An Hour Before Daylight: Memoirs of a Rural Boyhood*. New York: Simon and Schuster, 2001.

Carter, Jimmy. *Sharing Good Times*. New York: Simon and Schuster, 2004.

Chapin, Miles. *88 Keys: The Making of a Steinway Piano*. New York: Clarkson N. Potter, Inc., 1997.

Ciresa, Piera, Fabio Ognibeni, and Alessandro Tossani. *The Soul of the Wood*. Val di Fiemme, Italy: E. Ciresa S.R.L, 2002.

Consorzio Liutai & Archettai. ". . . And They Made Violins in Cremona," Cremona, Italy: Consorzio Liutai & Archettai, 2000.

Constantine, Albert. *Know Your Woods*. Guilford, CT: The Lyons Press, 2005.

Cook, Mary. *Loretto: The Sisters and Their Santa Fe Chapel*. Santa Fe: Museum of New Mexico Press, 2002.

Donatelli, Carlo. *The Gondola: An Extraordinary Naval Architecture*. Venice: Arsendale Editrice, 1990.

Ennos, Roland. *Trees.* Washington, D.C.: Smithsonian Institution Press, 2001.

Farb, Peter. *The Forest.* New York: Time-Life Books, 1961.

Fisher, Jim. *The Lindbergh Case.* New Brunswick, NJ: Rutgers University Press, 1987.

Flynn, James, and Charles Holder. *A Guide to Useful Woods of the World.* 2nd ed. Madison, WI: Forest Products Society, 2001.

Frazer, Sir James George. *The Golden Bough.* New York: The MacMillan Co., 1922.

Freese, Barbara. *Coal: A Human History.* New York: Penguin, 2004.

Hageneder, Fred. *The Meaning of Trees.* San Francisco: Chronicle Books, 2005.

Geiringer, Karl. *Instruments in the History of Western Music.* New York: Oxford University Press, 1978.

Gore, Al. *An Inconvenient Truth.* Emmaus, PA: Rodale Press, 2006.

Green, Harvey. *Wood: Craft, Culture, History.* New York: Viking Press, 2006

Groeschen, Jessie. *The Art of Chainsaw Carving.* East Petersburg, PA: Fox Chapel Publishing, 2005.

Guinness World Records 2006. London: Guiness, 2005.

Gurstelle, William. *The Art of the Catapult.* Chicago: Chicago Review Press, 2004.

Gutman, Dan. *Banana Bats and Ding-Dong Balls.* New York: Macmillan, 1995.

Hardy, Robert. *Longbow: A Social and Military History.* Goldthwaite, TX: Bois d'Arc Press, 1992.

Harlow, William. *Inside Wood: Masterpiece of Nature.* Washington, DC: American Forestry Association, 1970.

Hart, Clive. *Kites: An Historical Survey.* New York: Frederick A. Praeger, 1967.

Heinrich, Bernd. *The Trees in My Forest.* New York: HarperCollins, 1997.

Hill, Bob. *Crack of the Bat: The Louisville Slugger Story.* Champaign, IL: Sports Publishing, LLC, 2002.

Hoadley, Bruce. *Identifying Wood.* Newtown, CT: Taunton Press, 1990.

———. *Understanding Wood: A Craftsman's Guide to Wood Technology.* Newtown, CT: Taunton Press, 2000

Hurley, Victor. *Arrows Against Steel.* New York: Mason/Charter, 1975.

International Book of Wood. New York: Simon and Schuster, 1976.

Johnson, Duane. *How a House Works.* Pleasantville, NY: Readers Digest, 1994.

Keegan, John. A *History of Warfare.* New York: Vintage, 1994.

Kilby, Kenneth. *The Cooper and His Trade.* Fresno, CA: Linden Publishing, 1989.

King, Gil. *The Art of Golf Antiques.* Philadelphia: Courage Books, 2001.

Lauritzen, Peter. *Venice Preserved*. Bethesda, MD: Adler and Adler, 1986.

Lehane, Brendan. *The Power of Plants*. New York: McGraw-Hill, 1977.

Leslie, Edward. *Desperate Journeys, Abandoned Souls*. Boston: Houghton Mifflin, 1988.

Levy, Mattys, and Mario Salvadori. *Why Buildings Fall Down*. New York: W. W. Norton, 1994.

Lewington, Anna, and Edward Parker. *Ancient Trees: Trees That Live for a Thousand Years*. London: Collins and Brown Ltd, 1999.

Little, Charles. *The Dying of Trees*. New York: Penguin, 1997.

Logan, William Bryant. *Oak: The Frame of Civilization*. New York: W. W. Norton, 2005.

McRaven, Charles. *Building and Restoring the Hewn Log House*. Cincinnati, OH: Betterway Books, 1994.

Miller, Regis. *Xylaria at the Forest Products Laboratory, Past, Present, and Future*. USDA, Forest Products Laboratory, 1999.

Moldenke, Harold, and Alma Moldenke. *Plants of the Bible*. Waltham, MA: Chronica Botanica Company, 1952.

Moore, Patrick. *Green Spirit: Trees Are the Answer*. Vancouver: Green Spirit Enterprises, 2000.

Mumford, Louis. *Technics and Civilization*. New York: Harcourt, Brace and World, 1963.

Nakashima, George. *The Soul of a Tree*. New York: Kodansha International, 1981.

Nakashima, Mira. *Nature, Form & Spirit: The Life and Legacy of George Nakashima*. New York: Harry Abrams, 2003.

Nelson, Peter, Judy Nelson, and David Larkin. *The Treehouse Book*. New York: Universe Publishing, 2000.

Northern Cooperage. Inside *the Hoops: A Guide to Better Cooperage,* Brochure. St. Paul, MN: Undated.

Nowlin, William. *The Bark Covered House*. Charleston, SC: BiblioBazaar, 2006.

Oakeshott, R. Ewart. *The Archaeology of Weapons*. Mineola, NY: Dover Publications, 1996.

Pakenham, Thomas. *Remarkable Trees of the World*. New York: W. W. Norton, 2003.

Payne-Gallwey, Sir Ralph. *The Projectile-Throwing Engines of the Ancients*, 1906, 33. Reprinted in Barcia, Bernard: *Catapult Design, Construction and Competition*. Indianapolis, IN: Pompeiiana, Inc., 2006.

Peattie, Donald. *A Natural History of North American Trees*. Boston: Houghton Mifflin, 2007.

Peattie, Donald. *A Natural History of Trees*. Boston: Houghton Mifflin Co., 1948.

Pelham, David. *The Penguin Book of Kites*. New York: Penguin, 1976.

Perlin, John. *A Forest Journey: The Story of Wood and Civilization*. Woodstock, VT: The Countryman Press, 1989.

Petroski, Henry. *The Pencil: A History of Design and Circumstance*. New York: Knopf, 1992.

Plotnik, Arthur. *The Urban Tree Book*. New York: Three Rivers Press, 2000.

Porter, Terry. *Wood: Identification and Use*. East Sussex, UK: Guild of Master Craftsman Publications, Ltd, 2006.

Russell, Tony. *The New Encyclopedia of American Trees*. London: Hermes House, 2005.

Rutherford, Scott. *The American Roller Coaster*. St. Paul, MN: MBI Publishing Co., 2001.

Rybczynski, Witold. *One Good Turn: A Natural History of the Screwdriver and the Screw*. New York: Scribner, 2001.

Salvadori, Mario. *Why Buildings Stand Up*. New York: W. W. Norton, 2002.

Schwartz, Milton, and Robert Maguglin. *The Howard Hughes Flying Boat*. Oakland, CA: WCO Port Properties, 1983.

Silverstein, Shel. *The Giving Tree*. New York: HarperCollins, 1964.

Sloane, Eric. *A Reverence for Wood*. New York: Wilfred Funk, Inc., 1965.

Staten, Vince. *Did Monkeys Invent the Monkey Wrench?* New York: Touchstone Books, 1996.

Taylor, Troy. "The Winchester Mystery House," an overview of "The Winchester Mansion Story," in *Haunting of America*. Whitechapel Productions, 2001. (*Note*: This was online at www.prairieghosts.com/winchester.html. It explains it as "an overview of the Winchester Mansion Story. A more complete version of the events are chronicled in Troy Taylor's book, *Haunting of America*!")

Thiede, Carsten, and Matthew d'Ancona. *The Quest for the True Cross*. New York: Palgrave Macmillan, 2000.

Bonometti, Pietr. *Cremona: A Town Worth Discovering*. Modena, Italy: ITALCARDS Editions, 1988.

Visser. *Rituals of Dinner*. New York: Grove Weidenfeld, 1991.

Vittoria, Eugenio. *The Gondolier and His Gondola*. Venice: Editrice Evi, 1979.

Volkman, Ernest. *Science Goes to War*. New York: John Wiley, 2002.

Walker, Aidan. *The Encyclopedia of Wood*. New York: Quarto Publishing, 1989.

Williams, Michael. *Deforesting the Earth*. Chicago: University of Chicago Press, 2006.

MAGAZINES, JOURNALS, NEWSPAPERS

"A Billion Trees?" *American Forest,* winter 2007, 20.

Ashford, Kate. "The View From Above." *American Forests,* summer 2005, 30.

Barkman, Lars. *Preserving the Wasa.* The Forbes Prize Lecture, Stockholm, Sweden, June 4, 1975.

Batcha, Becky. "This Case Never Closes." *New York Times*, June 22, 2003, http://www.lindberghkidnappinghoax.com/brace.html.

Beals, Harold, and Terry Davis. "Figure in Wood: An Illustrated Review." Bulletin 486, Alabama Agricultural Experiment Station, 1977.

Betts, H. S. "Hickory and Golf." U.S. Forest Service, USGA publication, August 16, 1922.

Brown, Nicholas. "Meet Clam's Cousin from Hell." *Portsmouth Herald*, July 26, 2002, http://www.seacoastonline.com/2002news/07262002/col_wate/15908.htm.

Choi, Charles. "Secrets of the Stradivarius: An Interview with Joseph Nagyvary." *Scientific American*, June 10, 2002.

Christiansen, Donna. "The Ladder Link." *Forests and People*, Fourth Quarter, 1977.

Cooper, Simon. "Global Timber Smugglers—and How You Can Stop Them." *Popular Mechanics,* May 2007, at www.popularmechanics.com/science/earth/4215504.html?page=2.

Davidson, Brian. "Miracle Remembered." *Andover Townsman*, April 6, 2006.

DeWitt, Ron. "A Favorite Wood." *World of Wood,* Volume 60, Number 2.

Easley, Forest N. "A Stairway from Heaven? Loretto Chapel Stairway Wood Analysis and Technical Description." December 28, 1996.

"Engineered Wood and the Environment: Facts and Figures," APA—The Engineered Wood Association, 2006, www.apawood.org/level_b.cfm?content=srv_env_facts.

Fisher, Marshall. "The Feel of Wood." *Atlantic Monthly,* July 1995.

Ferruci, Maria Silvia, and Pedro Acevedo-Rodriquez. "Three New Species of Serjania (Sapindaceae) from South America." *Systematic Botany* 30 (2005): 1.

Fusco, Chris. "Blind Woodworker Crafts Lifelong Dream." *Chicago Sun-Times*, February 19, 2002, www.highbeam.com.

Giorgi, Sebastiano. "Forcole E Remi, Squerie Paline" (trans. Abbey Mahin). *Voga Venexia.*

———. "Il lavoro del paio" (trans. Abbey Mahin). *Arte Navale.*

"Wooden Soldiers." *Golf Journal,* July 2000.

Goldsmith, Belinda. "Speed Is Not King for All Roller Coaster Fans." *Boston Globe*, June 7, 2006.

Grace, Jim. "The Enduring Osage Orange." Missouri Conservationist online, www.mdc.mo.gov/conmag/1995/11/06.html.

Grauer, Neil. "Heavenly Strings." www.cigaraficionado.com, Winter 1995.

The Gristmill, December 2006, What's It.

Hall, William, and Hu Maxwell. "Uses of Commercial Woods of the United States. II. Pines." Forest Service Bulletin 99, 1911, as quoted in Peattie, Donald. *A Natural History of Trees*. Boston: Houghton Mifflin. 1948:12.

Hair, D. *Historical Forestry Statistics of the United States*. USDA Forest Service Station Bulletin No. 228, 1958.

Halvorson, Donna, Karen Yuoso, and Jim Buchta. *Minneapolis Star Tribune* special report. "Owners of Newer Homes Face Water Damage Risk." www.stocorp.com/PR.nsf/5d5efb09f77373618525666a004bca57/d4f316cb97c674368525692d006d00ea?OpenDocument.

Hope, Bradley."Archaeologist Finds Pottery, Wood Water Mains Downtown." *New York Sun*, November 27, 2006.

Hubbell, Sue. "Let Us Now Praise the Romantic, Artful, Versatile Toothpick." *Smithsonian*, January 1997, 77.

Huit, Katherine (Director of Collections, Evergreen Aviation Museum). "Wood Construction and Exterior Finish of the Hughes Flying Boat." (One-page handout from Evergreen Aviation Museum.)

The International Book of Wood. Ed: M. Bramwell. New York: Simon and Schuster, 1976.

"International Evaluation of the Preserve the *Vasa* Project," Stockholm, December 2006, http://www.vasamuseet.se/upload/bevara_vasa_rapport.pdf.

Jeffries, Tom. "Metabolic Engineering of Yeasts for Ethanol Production from Biorefinery Hydrolysates," http://www.metabolicengineering.gov/me2005/tjeffries.html.

"Joint DOE, USDA Grant Funds Development of Poplar Trees Optimized for Ethanol Feedstock," October 26, 2006, as posted on www.greencarcongress.com.

Lacer, Alan. "Spalted Wood." *American Woodturner* as posted at www.woodturninglearn.net/articles/spaltedwood.htm.

Lake, Morris. "The Magic of Huon Pine." *World of Wood*, September/October 2006, 30.

McCluskey, Mark. "Tennis Swaps Grace for Strength." *Wired News*, August 26, 2003, http://www.wired.com/news/technology/1,60177-1.html.

McNeil, Donald, Jr. "Yet Another Plague for Cursed, Glorious Ship." *New York Times*, May 14, 2002.

Mitchell, Gordon. "In the Work Shop Without Vision," www.woodcentral.com.

Mitchell, Gordon. "Vision for a Challenge," www.woodcraftmagazine.com.

Monteux, Pierre. Quoted in program notes, Lincoln Center playbill, January 2007, for "The Rite of Spring for Two Pianos."

Nowak, David J. *Benefits of Community Trees.* Brooklyn Trees, USDA Forest Service General Technical Report.

O'Brien, Michael. *The Five Ages of Wood.* Virginia Tech, SWST Conference address, June 2000.

Oo, Pauline, and Jeff Falk. "Fast-Growing Trees for Fuel and Fiber." May 26, 2006, UMN News. http://www1.umn.edu/umnnews/Feature_Stories/Fastgrowing_trees_for_fuel_and_fiber.html.

Pacher, Sara. "The Restoration of Jimmy Carter." *Mother Earth News,* November/December 1987.

Potter, Jerry. "Move to Metal Transformed Tour." *USA Today,* January 8, 2004, http://www.usatoday.com/sports/golf/pga/2004-01-08-metal-woods_x.htm.

Powell, Benjamin O. Letter regarding CPSC No. CP03-02, Petition for Performance Standards for Table Saws. Consumer Product Safety Commission Web site.

Preston, Richard. "Climbing the Redwoods." *New Yorker,* February 14 and 21, 2005, 212.

Reid, Joseph. "Wood Used in Violins." *World of Wood,* Volume 60, No. 1, 10.

Richie, Brooke. CBS, Dallas/Fort Worth affiliate, June 26, 2006.

Royal Navy Museum. *John Harrison and the Finding of Longitude.* Information Sheet No. 83, www.royalnavalmuseum.org/info_sheets_john_harrison.htm.

Scott, Charlie. "The Day the Spruce Goose Flew." *News Register,* McMinnville, Oregon, November 4, 1997, http://www.newsregister.com/ss/goose/Staff-Coverage/TheDayItFlew_110497.html.

Snyder, Michael. "How Is Birdseye Maple Formed?" *Northern Woodlands,* autumn 1999, http://www.mapleinfo.org/htm/bird.cfm.

Tsutsui, William. "Landscapes in the Dark Valley: Toward an Environmental History of Wartime Japan." *Environmental History,* April 2003, www.historycooperative.org/journals/eh/8.2/tsutsui.html.

USDA Forest Service Pamphlet #R1-92-100, www.coloradotrees.org/benefits.htm#carbon.

"The Virginian Pilot," August 4, 1995, http://scholar.lib.vt.edu/VA-news/VA-Pilot/issues/1995/vp950804/08040464.htm.

Weisse, Daniel. "Architectural Symbolism and the Decoration of the Ste.-Chapelle." *The Art Bulletin,* June 1, 1995, HighBeam Research Web site, http://www.highbeam.com/library/docfree.asp?DOCID=1G1:17239640&ctrlInfo=Round20%3AMode20d%3ADocG%3AResult&ao.

White, Bob. "Man Lifting Kites—An Overview of Historical Accounts." Port Colborne, Ontario, Canada, compiled and e-mailed to author May 25, 2006.

Wong, Margaret. "High-Rise Bamboo Scaffolds Can Make for a Deadly Climb." *Milwaukee Journal Sentinel,* October 22, 2000 (Associated Press).

"World's Largest Wooden Barrel Factory Flourishes in Missouri." *Modern Brewery Age,* September 18, 2000 (Associated Press).

"World's Oldest Building Discovered." BBC News, March 1, 2000, http://news.bbc.co.uk/1/hi/sci/tech/662794.stm.

"Vision for a Challenge: Gordon Mitchell." *Woodcraft,* July 2005.

"Wooden Water Pipe." *The Manufacturer and Builder,* Volume 18, No. 1. January 1886, 4–5, www.sewerhistory.org.

Yoder, Eric. Interview with Elmore Just, "Surprise: Persimmon Compares Well to Metal," www.golf.com.

Youngs, Dr. Robert L. "Every Age, the Age of Wood." *Interdisciplinary Science Reviews*, Volume 7, No. 3, 1982.

TELEVISON, RADIO, AND MISCELLANEOUS

"Ancient Discoveries—Machines I and II." History Channel, aired December 12, 2005.

"Stradivari." CBS News *60 Minutes,* Volume XXVII, No. 47, produced by John Tiffin.

"Leonardo's Dream Machines." PBS, aired February 2006.

"The Lumberyard." *Modern Marvels,* The History Channel, aired March 27, 2007.

Minnesota Department of Natural Resources Web site, "Reasons to plant trees," http://www.dnr.state.mn.us/forestry/nurseries/reasons.html.

Gerald Tuskan, radio interview, "Tree Research Could Lead to New Fuel Sources." *Talk of the Nation,* September 15, 2006.

PHOTOGRAPHY AND ILLUSTRATION CREDITS

Raimund Aichbauer: photograph, page 64
Ancient Wood: photographs, pages 4 and 5
Ark van Noach: photograph, page 304
Scott Beal of Laughing Squid: photograph, page 114
Bensonwood Company: photograph, page 33
Thomas Boehm/Ancient Archery: photograph, page 270
Allan Boardman: photograph, page 66
R.T. Bullard Plastering and Stucco: photograph, page 201
Spike Carlsen: photographs, pages 18, 37, 47, 74, 78, 97, 118, 123, 126, 146, 157, 179, 183, 214, 222, 230, 232, 272, 309, 318, 328, 344, 357
Carter Center: photograph, page 84
Ron Faulkner: photograph, page 61
Scotty Fulton: photograph, page 102
Tom Gibson: photograph, page 187
Michele Hermansen: photograph, page 352
Eric Jackson: photograph, page 169
Kayila Lodge: photograph, page 189
Louisville Golf: photograph, page 167
Livio De Marchi and Sam Barcroft: photographs, pages 54 and 55
Gordon Mitchell: photograph, page 58
McMullin Redwood: photograph, page 29
Johannes Michelsen: photograph, page 95
Stan Munro/Toothpick City: photograph, page 250
Musée du Louvre: photograph, page 92 (public domain)
Ray Murphy: photograph, page 52
Portugese Cork Association: photograph, page 14
Public domain: illustration, page 261
Public domain: photograph, pages 267, 290, 296, 301, 338
SawStop: photograph, page 105
Sewerhistory.org: photograph, page 322
Steinway Pianos: photograph, page 134
Karen Sun: photograph, page 25
Sam Talarico/Talarico Hardwoods: photograph, page 9

Malcolm Tibbetts: photograph, page 98
Tom Turnquist: photograph, page 110
Arnot Wadsworth: photograph, page 173
Warwick Castle: photograph, page 260
Wikipedia (free license): photographs, pages 130 and 204
This Wikipedia and Wikimedia Commons image is from the user Chris 73 and is freely available at http://commons.wikimedia.org/wiki/Image:Giant_Taiko_Drum_Nagoya.jpg under the creative commons cc-by-sa 2.5 license

INDEX

Page numbers in *italics* indicate illustrations.

ABOUT THE AUTHOR

Kat Carlsen

Spike Carlsen is an editor, author, carpenter, and woodworker, who has been immersed in the world of wood and woodworking for over 30 years. He is the former executive editor of *Family Handyman* magazine where he wrote hundreds of articles on home improvement and oversaw the creation of dozens of books, including the revised *Reader's Digest Complete Do-It-Yourself Manual.*

He has written articles for *Old House Journal, Fine Homebuilding, Workbench, The Minneapolis Star Tribune*, and other publications. He currently writes "Ask Spike" for *Backyard Living* magazine. He has made appearances on the CBS *Early Show, The Weekend Today Show*, WGN-TV, *Good Morning Texas*, HGTV's "25 Biggest Remodeling Mistakes" special, USA Radio, and many other national radio and television shows.

Prior to becoming an editor, he worked as a carpenter for fifteen years and ran his own construction and remodeling company, working on projects ranging from energy efficient homes to historic restorations.

He and his wife, Kat, have five adult children and live in Stillwater, Minnesota. He recently returned from Tanzania where he helped bring electricity to a remote secondary school, and installed a smokeless, wood-burning "rocket stove" for cooking. In his spare time he enjoys biking, restoring vintage radios, woodworking, and renovating (and renovating and renovating) their 1850s Greek Revival home.